Under the
Indian Ocean

Al J. Venter

Under the Indian Ocean

Nautical

Nautical Publishing Company Limited

First Published in Great Britain by
NAUTICAL PUBLISHING COMPANY LIMITED
Nautical House, Lymington, Hampshire SO4 9BA

ISBN 0 245 52020. 1

Other works by the author

The Terror Fighters: Profile of Guerrilla War in Southern
Africa (Purnells)
Underwater Africa (Purnells)
Underwater Seychelles
Portugal's Guerrilla War (John Malherbe)
War in Africa (Human & Rousseau)
Coloured: A Profile of Two Million South Africans (Purnells)
Report on Portugal's War in Guine-Bissau (California
Institute of Technology)
Assignment Africa (John Malherbe)
Africa Handbook (Human & Rousseau)
Editor of Underwater Handbook (Atlantic Underwater Club)

Made and Printed in Great Britain by
Cox & Wyman Limited, London, Fakenham and Reading

DEDICATION

TO

JAMES LEONARD BRIERLEY SMITH –
known to all his friends and students as J.L.B. – a
pioneer in his field across the vast expanse of the
Indian Ocean.

Acknowledgements

Without the support of many people in countries spread across four continents, this book would never have been completed. In producing a work about so vast an area as the Indian Ocean it has been necessary for me to call on various individuals as well as private and public bodies for help. I am particularly obliged to Mrs. Margaret Smith, widow of the late Professor J. L. B. Smith and present director of the J. L. B. Smith Institute of Ichthyology; Mr. George Hughes of the Oceanographic Research Institute, Durban; Dr. and Mrs. James Kirkman, warden of Fort Jesus Museum, Mombasa, and the International Oceanographic Foundation, Miami, for their assistance over the past year.

I am grateful too that Mr. Tom Bulpin has allowed me to use an extract from Professor J. L. B. Smith's book *High Tide*, published in South Africa in 1968. The map of Aldabra comes from Doug Alexander's latest book *Holiday in the Seychelles* published by Purnells, Cape Town, during 1972 while the map of the Indian Ocean was drawn by Dick Marra and reproduced with the kind permission of the International Oceanographic Institute, Florida.

Among the photographers who contributed to these pages are such notables as my very good friends Peter Saw, erstwhile Cousteau cameraman Gary Haselau and prizewinner Johnny van der Walt. Other photographers who are represented include Brian Rees, George Hughes, Peter le Lievre, Cliff Williams, Wolf Avni, Harry Barclay, Alex Pappayanni, Alex Fricke, Terry Attridge and Mike Clark. David Livingston of Tiburon, California, went to a great deal of trouble to provide excellent photographs besides his story on diving in the Red Sea. Seaphot of Britain provided a number of specialist photographs including those of black-tipped sharks taken off Ceylon by Arthur C. Clarke.

Among those who helped me during the course of my general ramblings while in search of material were my friends and co-divers Conway and Valerie Plough and Peter Hutchinson, all of Mombasa; the Saws of Malindi; 'Rafiki' Mohammed Amin of Nairobi; Chris Nicolas of Watamu, and Lallie Didham of Shimoni. In Dar es Salaam I was helped by John Navatta of Aquasports and Mike Reigels. Willy Halpert of Eilat did his 'thing' once more. Many thanks to you all.

My wife and fellow undersea-enthusiast played an invaluable role in helping me prepare the manuscript within the stipulated time in spite of other pressing commitments; as did Miss Sue Gompels and Miss Margaret Stegman, both of Cape Town. Louis Crole once more brought his 'eagle-eyes' to bear with the proofreading.

Cape Town
June, 1973.   *Al J. Venter*

Contents

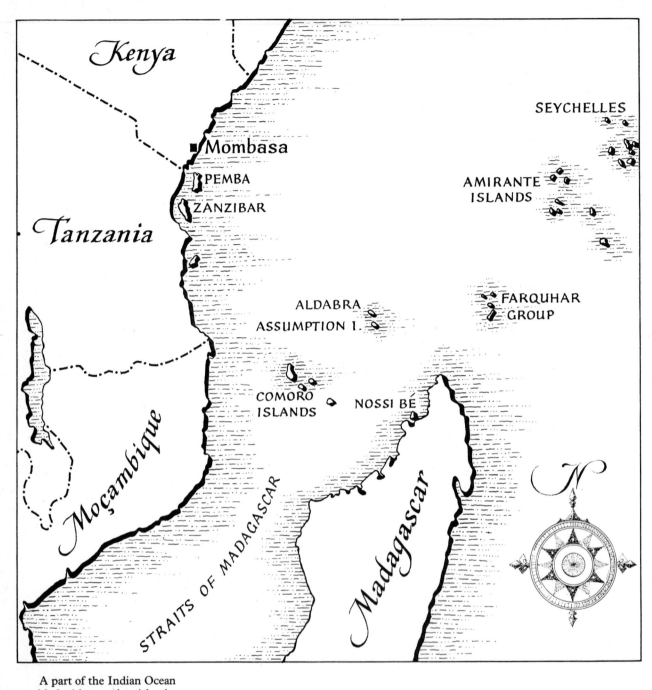

A part of the Indian Ocean
studded with countless islands,
and bounded to the west by
the coastline of East Africa in
which long stretches have no
harbours or facilities of any
kind for vessels.

The Indian Ocean

In the modern period it was the late Professor J. L. B. Smith who often referred to that vast expanse of sea bounded by Africa in the west and by Australia and the Indian sub-continent in the east as 'The mysterious Ocean'. He and his wife Margaret explored this Ocean's limits for the best part of a lifetime and the old savant regarded aspects of this vast water mass with awe and respect until the day he died. In 1960 he wrote: 'There, clearly, much awaits discovery – I venture to predict that in addition to numerous material benefits to the nations in that part of the globe there may be more spectacular discoveries in the Indian Ocean, such as further nests of coelacanths, or even other unexpected archaic survivors, and almost certainly, one or more of the fabulous sea-serpents that undoubtedly exist in all oceans.'

It always remained Professor Smith's view that while all parts of the land had been explored and conquered, many of the inner secrets of the vast oceans still remained largely unknown. Only in the sixties was the first constructive step taken to solve some of the enigmas that have stared man in the face for millennia. In addition to this, he said, what little research had been done had been limited to the Mediterranean besides the Atlantic, and Pacific Oceans. The Indian Ocean, including the western Indian Ocean – that borders on much of Africa – and which forms the basis of this book, still largely remains the most poorly known region of the Earth's surface.

This vast body of water, one-seventh of the total surface of the Earth has an area close on thirty million square miles. It has certain features not found in any other part of the world, the most striking being the manner in which the prevailing winds reverse their courses with the seasons.

There are many reasons why this ocean is so little known. The use made of any sea for transport depends largely on what lies ashore – densely populated developed countries have heavy sea traffic. East Africa, for example, is almost unique in its present condition. It has one of the longest single coastlines in the world and a teeming population, but most are relatively undeveloped people and their countries are struggling to keep pace with progress in the Northern Hemisphere. There are, in consequence, long stretches of the coast in East Africa (and Madagascar, for that matter) without proper harbours or facilities of any kind for vessels. Most sea traffic in these parts is utterly remote, almost quite cut off from the peoples of the shores; those passing in ships, near, for instance, the Rovuma area of Tanzania and Northern Moçambique are virtually in another world, so great is the contrast between the wild and primitive conditions ashore and those on a modern liner.

The Indian Ocean is also a region which is comparatively sparsely populated. While places like Ceylon, India and Mauritius have their teeming millions the majority of islands and archipelagos rarely boast impressive population statistics; and these speckles of black on a map are few and far between. It was only a few years ago that an American alone in a small yacht was stranded on an island somewhere to the west of the Australian coast. He made himself a primitive boat and drifted across most of the thousands of miles of the southern-central Indian Ocean without seeing, or being seen by a single other craft. He was finally cast ashore at a tiny island near Alphonse Island where he was rescued, in extremity, only by chance.

Another group of scientists spent two months in an area just north of Madagascar. During their entire research period – and this happened during the sixties – they never saw another sail or ship. To all intents and purposes they could have been castaways at the end of the world.

But it was marine life that originally brought Dr. J. L. B. Smith to travel and explore the Indian Ocean and to any undersea enthusiast who has not yet experienced the wealth and variety of an Indian Ocean dive, much still lies in store. Unlike the Mediterranean and Caribbean seas which have been all but cleared of any large shoals of gamefish, the Indian Ocean still has the greatest variety of marine creatures of any ocean; large and small. And they are there in great quantity.

Professor Smith predicted that proper investigation may yet find that even those parts where the present fish supply is poor may well hold hitherto undreamt of resources of valuable food fishes. That this is not unlikely was shown when in July, 1957, ships crossing south of the Bay of Bengal not far from India reported fish mortality on a fantastic scale. The surface of the ocean for about 100 000 square miles was covered with dead fish. It was surmised at the time that this was due to a diversion of some cold current, which is not uncommon in the seas; if the water is cold enough, fish die in their millions. A Russian scientist who was in the area calculated that the fish killed in that stretch of ocean, whatever the cause may have been, *exceeded the total world catch of fish for a whole year.* What is more interesting about this phenomenon is that this happened in an area where fish are normally regarded as scarce. Similar but smaller marine catastrophes occurred about the same time to the north-west, near the Persian Gulf.

What is abundantly clear is that the Indian Ocean on its own is an area of fascinating contrast, and offers more than a touch of the mystique. Those who have visited the zone have never failed to be enchanted. A single dive may bring a brief adventure with giant manta rays or a mammoth whale shark. It could also provide visual impact among some of the most beautiful coral reefs to be found anywhere. And, with improved communications, this remote region is gradually being opened up still further.

In the final reckoning it is man who will control the destiny of the Indian Ocean. He will enjoy what he sees, but the question must be asked: will he protect this heritage? One can only hope that a lesson has been learnt from other quarters of the globe where marine life had been indiscriminately plundered and that what has been left, polluted; for posterity to ponder why.

The large Indian Ocean
grouper or rock cod which
can grow to as large as
2 000 lbs off the East Africa
and Madagascar coasts.
(Peter Saw)

Chapter 1

Shipwrecks of the Seychelles

The Seychelles Islands sprang into world prominence with the wreck of the Royal Naval Fleet Auxiliary Ennerdale in June, 1970. Once the haunt of East Indiamen and privateers there are many other wrecks lying off the reefs of this ocean paradise which General Gordon once believed was part of the original Garden of Eden.

As wrecks go, the Royal Fleet Auxiliary *Ennerdale* does not lie deep. Her ragged, twisted hull reaches upwards from the ocean floor to within about thirty feet of the surface.

Visibility on the reef on which this 47 000-ton tanker disembowelled herself is usually good; from the surface it is possible to see much of the ship as she rests on her side with her stern gently receding into deeper water. Lying only eight miles to the north-east of Victoria, the position is reached in an hour. The fun really starts once lungs are strapped on and you drop overboard. Apart from the jagged holes – she was blasted by Royal Navy divers within days of her grounding in June, 1970, to disperse the oil still in her tanks – the tanker is in good shape. Little marine growth has settled on her steel plates and the dull grey navy paint reflects a neutral tint on the clear blue waters around her.

With the right equipment it is possible to explore much of the ship where she lies. Grouper, rock cod, and an occasional moray peer out apprehensively from the cavernous depths they now regard as home, giving the passing diver more than casual attention. Down aft, much of the accommodation area still lies intact, as does part of the engine-room, offering tempting if somewhat bizarre insight through a number of half-open doors, hatches and gaping jagged holes. It takes a brave man to go in, for the black interior is mysterious and uninviting as it will continue to be until this vast steel hull slowly disintegrates in the face of encroaching corrosion and the occasional storm which sweeps across the area in a flurry of white water and foam.

Certainly the most hazardous task after the *Ennerdale* had grounded was placing explosives in order to breach the remaining tanks. A favourable swell was running at the time which promised to disperse the remaining oil in open waters away from the island. No time could be lost, but because of heavy seas divers could not be used for much of the operation.

Lieutenant-Commander Braidwood, Far East Clearance Diving Officer, working with Lieutenant Kenworthy, senior pilot detached from 847 Naval Air Squadron, devised a system whereby three mortar bombs placed on a pallet could be lowered from a helicopter into position on the wreck; the cordtex fuse was then ignited by the

Royal Fleet Auxiliary *Ennerdale* in Cape Town on her final voyage to Mahé where she sank. (*The Argus*)

crew inside the 'chopper'. These bombs effectively breached the port tanks, but to reach the starboard tanks underneath this mass of twisted steel, another demolition charge was constructed. Divers were obliged to enter the sea and secure a forty-five-foot wire pennant to the top of the wreck and using a Gemini inflatable dinghy, joined it to a similar wire lowered from the helicopter. The dinghy raced clear, the helicopter crew lit the cordtex fuse and released the bomb tray which, on its ninety-foot pennant sank down alongside the vents on the lower tanks. The blast provided a mighty spectacle.

Later in the operation as H.M.S. *Cachalot* was unable to fire torpedoes into the wreckage to free oil trapped in the hull, the warheads were taken off the torpedoes and lowered to be detonated in the same way. When some of the warheads failed to explode, Lieutenant-Commander Braidwood dived to investigate the reason.

At this stage short pre-cut delay fuses were added and Lieutenant-Commander Braidwood ignited fuses on the surface of the sea while suspended from the helicopter's winch wire. There was always a swell, often heavy, and entering the water so close to a large wreck in this kind of weather was dangerous. Faced with a difficult and unusual situation Lieutenant-Commander Braidwood certainly showed remarkable ingenuity and courage.

But the *Ennerdale* is not the only wreck lying in Seychelles waters. Straddling one of the sea routes to India, these waters and their mysterious reefs have claimed numerous ships over the past four centuries, from the French frigate *Heureuse* which in 1763 was wrecked with treasure onboard on the reef off Providence Island in the Farquhar group, to the *Lord of the Isles*, a coaster which was lost in 1906 on the north coast of Mahé without any trace of wreckage. Some say her jagged old spars have been found; others dispute this claim but the *Lord* provides interesting speculation for some of the more venturesome divers who visit the island.

Even for those who do not take their sport very seriously, Victoria Harbour can provide the lure of sunken treasure. A Portuguese ship, the *Parachi Pachia* sank within the precincts of the port in

Decompression after a dive.
(Colin Doeg: Seaphot)

August, 1801, during a battle between the English and French frigates *Sybille* and *La Chiffone*. Sixty years later, the *Voyageur*, under Captain Arnold Loizeau, also sank in Victoria Harbour. She was burnt out and most of her cargo was lost.

Further afield there are wrecks on Platte Island, south of Mahé, and on Denis Island to the north. Platte Island claimed two ships; the French Government corvette *La Perle* in May, 1863 and a lugger of forty-five tons, the *Alice Adeline* which grounded in 1906. Off Bird Island the whaler *Greenwich* struck a reef in February, 1833; some of her old coral-encrusted ribs are still to be seen along the outer reef area.

One of the most interesting wrecks to occur in this stretch of the Indian Ocean was the loss of the *St. Abbs*, a British ship on passage from London to Bombay. One dark stormy night in June, 1854, the *St. Abbs* went aground on one of the reefs off the Farquhar Islands, lying just south of the Amirantes. Only six people finally reached shore; the rest drowned or were taken by a number of large sharks attracted to the scene of the disaster by stores which had been thrown overboard to lighten the vessel.

One of the passengers onboard the ship at the time was a young cadet, Edward Ross, on his way to take up a post in India. He described the incident vividly in a report which is still held by the Seychelles Government archives:

My friend and comrade Bell concurred with me that our only chance was an attempt to swim ashore, and we at once prepared to carry out this resolve. I attired myself in a shirt and white trousers, with a flannel vest and flannel drawers underneath. Round my waist I tied in a handkerchief a tin of canned milk, a pipe and some tobacco, removing shoes and socks. Before leaving the wreck I went to say farewell to Mr. Stone, the mate, who was still in a low state of despair. Poor man, no wonder! He had a wife and family and, being unable to swim, his case was well-nigh hopeless. I tried to cheer him all I could, saying that Bell and I would reach the shore and help in the matter of the spar and line. And so with a goodbye to all we left on board. I slipped down a rope into the surf followed closely by Bell. When one is braced for the peril of such an attempt, the danger of it is lost from the mind, and the thoughts are bent on straining every nerve to succeed, and it is a good one. If, moreover, a man has a thought of serving others besides himself, there rises a feeling of exaltation, almost elation, which has probably carried many through imminent peril. All the same, had I known that besides the danger of the surf and breakers, there was a still more appalling one to run, I know not if I should have had the courage to face it. For afterwards I discovered that the surf swarmed with sharks. As soon as Bell and I reached the water, a heavy roller overtook us, and parted us for ever. I was told afterwards that, while apparently swimming through the surf, he was seen by those on the wreck to throw up his arms and suddenly disappear, and I fear he must have been dragged under by a shark.

The six men stranded on the desolate spit of land were obliged to keep themselves alive on the flesh of raw birds. There was no water on the island and the castaways drank liquor and vinegar washed up from the wreck which only aggravated their thirst and their plight. Ross' story of what happened then makes vivid reading. He continues his tale:

I think we were about seventeen days on Bird Island when, all being propitious, we set out on our arduous enterprise wading over the coral reef and pushing along our raft to which had been secured all our precious stores. I had tied some cloth over my feet to protect them from the sharp coral, but still the walking and wading was very painful, and at times I was forced to cling on to the raft. Sometimes we had to cross deep pools, and then we found ourselves attended by an escort of sharks. However, they were timid, poor things, and a splash at their noses easily drove them away. By afternoon we gained the shore of Juan de Nuovo Island, and lay down to rest before exploring our new dominions. A search party then went forward and, to our inexpressible relief, found pools of muddy fresh water about halfway up the island, which was perhaps three miles long and half a mile in width, covered with brush-wood and stunted trees. Birds were numerous, roosting and nesting on the scrubby trees. Our raft was pushed along the shore through the shallow water, and next morning we reached the further extremity of the island, where there was an eminence, a stunted palm tree, and some wells of brackish but drinkable water. A mat hut too was there, and of course we decided to make that spot our camping ground. We also found a cage of turtle in the sea close by, with some live turtle in it. Thus our position and prospects were much improved, and altogether more hopeful.

Evidently this island was not wholly forsaken and unvisited. The first night of our sojourn thereon we were startled by the howling of what seemed to be wild beasts, and later we ascertained that there were some black dogs living in the scrub, become quite wild and savage. They apparently roamed along the shores by night, and picked up crabs and such things for their living. The carpenter was taken aback one day by meeting the dogs, and came back to camp looking very scared.

Our daily routine was now somewhat varied. The boobies, a larger kind of seafowl from those on Bird Island, sat on the bushes and we would sally forth each morning armed with sticks to obtain the necessary supply. The birds sat stupidly, and waited most complacently to be knocked on the head. They were, however, rather scarcer, and we began to fear the supply would soon be exhausted. The frigate or man-o-war bird was a denizen of the island. The Captain one day brought back one of these birds and, much to our amusement, ate it for his dinner. Judging by the odour, the boobies were evidently comparatively flavourless. At night we slept in the hut we had discovered. Perhaps our slumbers there would have been less agreeable had we known that it had once been the abode of lepers.

The weary days wore on, and though our diet was varied by an occasional catch of turtle, it began to tell on our health, and the desolate feeling of being cut off from the world was becoming unbearable. We even thought of launching out to sea on a raft to take our chance, rather than linger on in such a wretched life. Is life worth living, do you ask? Try a spell on a desert island with Hope for your companion, and answer.

Five weeks later Ross and his party were rescued by a passing ship and taken to Mahé. But the story does not end there. He was eventually taken to India and after his arrival there, Cadet Ross served in the Central India Field Force during the Indian Mutiny. He later became Consul-General for Fars and the Persian Gulf, was knighted in 1892 and died in 1913, aged seventy-seven.

Other ships which have been lost in the Farquhar group are the *Hardwick Castle* still lying on the reef after seventy years and another British ship, the *Aymestry*, which went down in 1897.

Immediately to the west of the Amirantes is Cosmoledo Island. Like Farquhar, this ring of coral has taken its toll of merchant shipping, one of which has never been found. These include a Norwegian bark *Hamengia*, a total loss in 1913, and the auxiliary schooner *Meredith A. White*, which went down little more than a decade later. In 1874 the *Merry Monarch* sank on Wizard Island, another of the Cosmoledo group.

The island which has claimed more wrecks than any other is in all probability Providence Island near the Farquhar Group; like all the others mentioned, it is part of the Seychelles Archipelago. Altogether eight wrecks are listed on this strip of beach and coral in the middle of the Indian Ocean, from the French frigate *Heureuse* to the S.S. *Syria*, which went down shortly before the last war and whose hulk is still visible at low tide. Other ships off Providence include the British brig *Aure* (1836); the French barque *Federation*

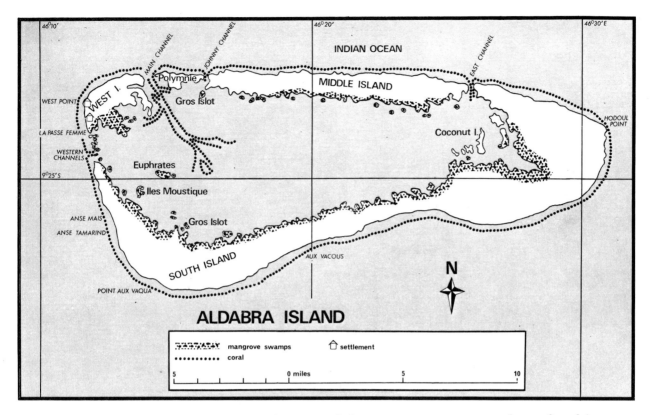

A dot on a tropical sea, Aldabra is the largest coral atoll in the world. It is 23 miles long and about eight miles wide. Unlike its neighbouring islands in the British Indian Ocean Territories, Aldabra's natural beauty has not been ruined by guano digging.

Lying about 250 miles north-west of Madagascar and 400 miles from the East African mainland this arid, rugged and unwelcoming atoll consists of four islands in a broken ring of dissected coral interrupted by tidal channels where the sea has broken through. The dunes of the south coast are exposed to fierce winds for most of the year.

The giant tortoises of Aldabra are now known to number about 100 000. When their relatives on surrounding islands were wiped out by raiding ships in search of food, the rugged atoll protected its own and apparently very few of them were taken.

The Marine life around the coast is rich and varied. There are huge shoals of humphead wrasse and giant parrotfish, schools of manta rays numbering over 200 and dolphins and turtles without number. Although the island is remote from all tourist haunts those who have dived off Aldabra maintain that the experience is unique and never to be repeated in any other ocean, mainly because the fish, protected by law, know few intruders apart from their natural enemies. The area has been declared a national park and is likely to remain so for the foreseeable future. Britain's Royal Society has financed a number of expeditions to the island.

(1894), the schooner *Maggie Low* which ran aground on a sand bank twenty miles south of Providence, also two ships which went down in 1906, the British *Endeavour* and a Norwegian barque *Jorgen Bank*.

Another ship to go aground in these parts was the Norwegian barque *Dagmar* in 1907. A tidy little craft of almost 400 tons she was on charter to the Mahé Syndicate Limited, and ran into trouble on her way to St. Pierre from Mahé. On the night of July 13 a heavy storm blew up and took the vessel some miles off course in treacherous waters. The chief mate who was on watch at the time said it was impossible to see or hear anything because of the wind and rain. The ship struck and although everything was done to get her off again, huge seas carried her further into the surf. All her crew and passengers were safely landed.

The Amirantes have also claimed a fair toll of tonnage. On Alphonse Island just south of the Amirantes three ships have been lost. A large French coal burner the S.S. *Dot* went down in November, 1873 and for half a century her funnel sticking out of the water acted as an unofficial warning to other shipping of the hazards of the reef. Parts of this ship are still to be found and although I have not dived there myself, word has it that the area is fascinating. Two other ships which came to grief off Alphonse were the *Tamatave*, a British craft of 196 tons in 1903, and the auxiliary schooner *Utopia*.

Of all the islands in the archipelago Astove is probably the one whose name has been consistently linked with treasure. Part of the Cosmoledo group, Astove lies just west of the Farquhars. Here, legend has it, the Portuguese frigate *Le Dom Royal*, laden with plunder and slaves from the East, went aground in 1760. Although almost everyone reached shore safely, the captain and crew attempted to reach Moçambique in a longboat and were never heard of again.

Other ships called and gradually word got about the Indian Ocean that a rich cargo of slaves was 'waiting to be taken' from the island. Two French ships set out from Mauritius to bring the slaves away and both were wrecked under mysterious circumstances on the same reef. Twenty years later another ship called and found the island inhabited solely by Blacks who 'on our approach to the shore set up wild shouts of defiance and placed themselves in an attitude of defence!' The vessel left again and reported the incident to the authorities at Reunion, then known as Bourbon. This time another vessel was fitted out for the expedition, but on her arrival at Astove she followed the fate of those which went before and was lost.

A British ship followed in 1796 from Mahé and attempted to take the slaves by force. Many of the Blacks and a few Europeans were killed in the fray which followed. Later another ship, having embarked about 100 slaves, was wrecked as she tried to negotiate a passage through the reef to her way to the open sea. All onboard were drowned or taken by sharks. The balance of the castaways were later embarked for the Seychelles leaving on the island only one man, a Black named Paul from Mahé. But he too disappeared mysteriously, for when a Captain Scoi passed Astove on his voyage from Mauritius to Moçambique in 1799 he found the place deserted.

Other islands in the Seychelles group that have wrecks which are often dived on by enthusiasts are as follows:

Aldabra :   S.S. *Glen Lyon* (1915).

Coetivy Island :   Corvette *Eclair* (1787).
                       Barque *Windsor Castle.*

Assumption Island : Lugger *Reve* (1915).

Desroches Island :   Barque *La Louise*, a slave trader which went down in 1809.

Remire Island :   British Sloop *Spitfire* (1801).

Chapter 2

Imperilled Gift of the Sea

Even in the far-flung reaches of the Indian Ocean the sea turtle is threatened with extinction. **George Hughes,** *long an authority on these friendly, if somewhat elusive creatures, tells of the life cycle and habits of the loggerhead turtle,* Caretta caretta, *a species which has been all but eradicated from many of its Indian Ocean haunts.*

'After loading some fifty turtles on deck we had an excellent repast of turtle meat and later prepared to sail with the tide'
(From the log of a French man-o'-war at Reunion Island, 1614).

Ship's logs of an age forgotten contained countless references to sea turtles on numerous islands and inlets of the Indian Ocean and judging from this quote, sea turtles played a prominent role in the history of early maritime exploration. The first recorded fight between Australian Aborigines and the crew of Captain Cook's historic voyage was over a pair of sea turtles which friendly Aborigines saw lying on the deck of the *Endeavour*. Knowing no traditions of ownership the Aborigines claimed them. They were restrained from doing so by the crew.

Sea turtles were caught in their hundreds on Ascension Island in the Atlantic. These helped to feed slaves carried to the Americas in transatlantic convoys. Other ships returning to London eagerly collected turtles for they fetched a very high price on the British market. At that time there appeared to be inexhaustible numbers of both slaves and turtles and indeed for many years both commodities were heavily exploited with consistently heavy losses. Local sources of supply were soon exhausted, so collectors started searching farther afield. The Indian Ocean provided a rich haul for those passing round the Cape on the eastward journey.

Slaughter has taken its toll and consequently Reunion and Mauritius today have strict laws to protect sea turtles that once swarmed ashore in their countless thousands to lay their eggs. But even there the strictures came too late and on both islands few turtles nest today. In Natal the first legislation enacted to prohibit the killing of nesting turtles was passed in 1916. But restrictions have hardly offset the natural balance.

The coastline of Natal is long and until relatively recently, the north-east coast, bordering an area known as Tongaland, was virtually undisturbed. Its people the *ama-Thonga* were unsophisticated and superstitious and by this fortunate set of circumstances not all the nesting populations of sea turtles were wiped out. Increasing contact with civilization, however, eroded many beliefs. Primitive fears which had protected the nesting turtles were overcome and earlier this century the Thongas started to kill turtles where they could find them.

The conservation programme had small beginnings. Twenty years ago an enterprising British scientist, Dr. Tom Harrison, started to take a deeper interest in the turtles nesting on some small islands near Kuching, Sarawak. Enthusiastically he wrote a number of papers which soon brought him into contact with the second turtle conservation pioneer, Dr. Archie Carr of Florida.

Their combined efforts set off a reaction which has led to the formation of an international commission to try and halt the sharp decline of sea turtle populations and, if possible, to try and build them up again. They have also stimulated a great deal of scientific research into the seven extant species of sea turtles: the green turtle *Chelonia mydas* (one of the world's most valuable reptiles and found in the Indian Ocean); the flatback turtle *Chelonia depressa*; the hawksbill turtle *Eretmochelys imbricata*; Kemp's ridley *Lepidochelys kempi*; the olive ridley *Lepidochelys olivacea*; the logger-

Hawksbill turtles, the source of all genuine tortoiseshell. (George Hughes)

head turtle *Caretta caretta* and lastly, the second largest reptile in the world today, the leatherback turtle *Dermochelys coriacea*.

Leatherback turtle nesting on the East African coastline. (George Hughes)

Local authorities such as the Natal Parks, or the Game and Fish Preservation Board in South Africa quickly responded to the discovery of sea turtle nesting populations, so in Tongaland they soon put a stop to all further killing by the Thongas. For a decade this body has exercised rigid control on the breeding beaches. Research has shown that the populations of loggerheads and leatherbacks are increasing and it appears that at least one battle in the war of conservation is being slowly won.

What makes the existence of any species of animal, least of all the rarely seen sea turtle, worthwhile saving?

Clearly, the sea turtle should be saved, firstly, out of gratitude, for its sorry state today results as a direct result of the majestic creature providing food, leather, oil and money to exploiters the world over.

One question often asked by the uninitiated is why the sea turtle should be respected.

The answer is simple. The sea turtle is a fellow creature entitled to his place on this earth as much as any human being, perhaps more so because it has a history some one hundred times longer than man. The turtle's present form has taken 200 million years to shape and its present life cycle and behaviour are just beginning to be understood.

In the rather humid atmosphere of the Permian age, some 200 million years ago, a strange creature was experimenting with its new mode of life. Executing some anatomical contortions, it somehow managed to arrange its girdles within its rib case with almost its whole body encased in a bony shell. Steering and mobility necessitated a halt at this point and if those ancient creatures had gastronomic characteristics anything like their present descendants, it went first in search of food. The armoured stomachs prospered and spread across continents. It had not entirely finished shaping its body and nature provided suitable shapes for living on land, in fresh water and finally for some of the more enterprising reptiles, a design for living in the sea.

Land living inflicted certain restrictions on size but water gave a little more leeway to the prototype beast as he then was. Consequently the sea turtle as now he became, attempted to match his dinosaurian relations by attaining staggering sizes. In the Cretaceous Age, 60–70 million years ago, the turtle had many forms, some of which measured two feet across the head and had carapaces over fifteen feet in length.

The fossil pages of the earth, however, are filled with stories of nature's grand experiments, many of which were not successful. Consequently numerous turtle extravaganzas blossomed and perished leaving only a few silicified bones to tell history of their passing.

And so, millions of years after turtles had started their struggle for a niche on earth, a certain happy balance was achieved. Turtles had prospered in the sea and there were millions of them. True, they were not without their enemies and many perished annually to keep other animals alive, but their reproductive rate more than adequately insured that these losses were not serious.

Sea turtles are not maternal. As a consequence no attending female waits protectively on a nesting beach for her eggs to hatch. After a breezy night in early February there is little on the long stretches of open shore in Tongaland to indicate the presence of 200 000 turtle eggs scattered in nests along thirty or more miles of beaches.

With very rare exceptions nearly all nests contain fertile eggs, normally of high percentage, in which the developing embryo is curled up with the head tucked between the hind flippers. Buried three feet under the surface the temperature of the eggs remains about three degrees centigrade higher than the surrounding sand which, acting as an insulator, effectively prevents a loss of heat generated by the developing eggs. Few enemies bother the buried clutches although occasional losses are experienced to ghost crabs (*Ocypode sp.*), red ants (*Dorylus sp.*), monitor lizards (*Varanus sp.*) and dogs.

Depending on weather conditions during the summer season the hatchlings take between fifty-five and sixty-five days to complete their development, after which they cut their way out of the egg by means of a sharp egg-tooth which is situated on the tip of the beak. At this point the eggshell, always soft even at laying, is brittle and tears easily. The hatchling turtle's first move on leaving the egg is to stretch out flat and adjust the soft bones and shell into the shape which will serve it throughout the years that lie ahead. While these alterations are made the umbilical cord through which the nutrients from the yolk were carried to the hatchling dries up and breaks off, freeing the hatchling for its first movements towards the sky it has never seen and the sea it has never experienced.

The hatchlings work in unison. As if at a signal, the entire batch of tiny animals start scrambling at the roof and walls of the egg chamber. This is a frenzied activity which requires and consumes considerable effort. Fortunately the hatchling occupies less space than it did as an egg and the resultant gaps soon fill with air filtering through the surrounding sand. Two hundred tiny green flippers soon bring the sand tumbling down from the roof; it passes through the thrashing mass and forms a new floor which gradually thickens carrying the hatchlings in a body towards the surface of the beach. Nearing the surface the hatchlings have been provided with an intrinsic protection mechanism which prevents them from emerging on to the Tongaland beaches in the middle of the day. If the sand surface temperature is too high there is inherent inhibition of movement or 'stalling action' and the hatchlings stop climbing. In this position they lie still until the temperature drops to a more tolerable level.

With the coming of nightfall, the sand temperature drops and with a final upward thrust the hatchlings burst on to the beach . . . a short pause for a look around . . . and again there is a concerted effort to reach the sea in as direct a line as possible.

But nature has not yet completed her natural selection process, so between the hatchlings and the sea are a variety of ghost crabs that help to remove the weaker of the breed.

Each hatchling should be equipped with the innate ability to distinguish low elevations and to judge, almost perfectly, the exact centre point of a relatively powerful light source, in this case; the sea horizon. The Tongaland beaches are backed on the landward side by high bush-covered sand dunes which at night are visible to the tiny hatchlings as an enormous black mass. In contrast, the sea almost always has a bright horizon which attracts the hatchlings in a frenzied rush.

From the original 120 eggs laid in the nest only about 90 hatchlings will get as far as the sea after the various selection gauntlets. Hurling themselves at the waves they are often washed up time and

Leatherhead hatchings in Tongaland. (George Hughes)

again, but turning themselves over (an ability not shared by their elders) they keep trying until eventually they are through the wash zone and on their way into the open sea where more dangers await them.

Having learnt to walk without guidance the hatchlings now swim immediately with the same facility. If they do not, the selection process continues to operate. Poor swimmers and imperfect breathers are soon picked off by fish attracted by their clumsy movements. Their imperfect genes are removed from the future gene pool of the loggerhead population.

Although considerable energy is expanded by the hatchlings over this period, the feeding mechanism has not yet become operative. At the primary stage the one common object of the hatchlings is to swim as far and as fast as possible away from the predator-infested coastal zone of Tongaland. Each hatchling has within it a small store of yolk which should be sufficient to last it from three to four days. Having no friends in the sea, the most that can be said for the marathon infant swim is that the farther these tiny creatures are from the coast the fewer enemies they have to contend with in the immediate vicinity. Relative safety is found only when they are surrounded by tremendous expanses of open sea. Never will they be free of all enemies; sharks will plague them for the rest of their lives.

The exact nature of the next stage of development of turtles is imperfectly understood because of the obvious difficulties in following a tiny creature, barely an inch and a half long, into the Indian Ocean. One thing that is certain, is that they swim directly out to sea. It is thought, however, that for many of the hatchlings the decision as to what direction to take is made for them by the Agulhas Current sweeping southwards through the Moçambique Channel from well to the north. Swimming into this swift, warm stream the hatchlings are soon borne southwards towards the tip of Africa.

Once their natural reserves have given out the tiny loggerheads start to look about for food, and here they are helped by the sea around them. Thousands of small floating organisms such as the Portuguese Man-o'-War (*Physalia sp.*), also known as the 'blue-bottle,' and many of its coelenterate cousins such as *Porpita* and *Vellela* provide ready meals. Floating molluscs such as the curious purple storm snail *Ianthina sp.*, and the 'flying' snails the *Pteropods*, are also taken. The latter two species not only supply much-needed protein but also calcium from the shells.

Having no teeth, the hatchling must tear soft portions from the prey animals. Nature has provided them with two small claws on the fore flipper to achieve purchase on the victim; once a good grip has been taken the hatchling jerks its head back and thrusts away

Green turtles' clutch of eggs
as they are laid on Casaurina
Island, Moçambique.
(George Hughes)

Green turtle that has
completed laying in North
Moçambique.
(George Hughes)

with the flippers, tearing off a portion. One of these claws soon becomes redundant and is lost long before the hatchling reaches adulthood.

With adequate food and warm temperatures the hatchling grows rapidly, adding a quarter inch to its overall length each week. A rapid growth-rate has high survival value for each day sees it beyond the capabilities of some predators. In the meantime the original group is dwindling day by day as more hatchlings fall prey to large marine animals.

Depending on prevailing weather conditions, strong winds can blow the hatchlings right over the current and out of its influence in the central ocean gyral south of Madagascar. Other winds may blow the hatchlings back out to the western side or perhaps ashore along the coast of Natal and the eastern province, or even down to Cape Agulhas at the southern tip of Africa or False Bay. It is from the study of the hatchlings washed ashore that possible movements have been assessed, as there is a tendency in the stranded specimens to be larger the farther south they are found.

Some hatchlings, having been in the fastest part of the current, achieve the voyage of nearly 1 000 miles to Cape Agulhas in a matter of two weeks, but most seem to take about two months.

It is interesting that after the first two-month period the ensuing few years of the loggerhead's life are largely hypothetical. They simply disappear from sight and do not reappear in home waters until they are between five and six years old. Of hundreds of turtle remains studied on the coasts of South Africa, Moçambique, Madagascar and the Mascarene Islands, not one has fitted into the age group of six months to five years. Juveniles of other species such as the green and hawksbills make their appearance as young as one year old, but not a single loggerhead has been found in home waters.

It has been put forward that not all loggerheads survive to lead an apparently idyllic existence of just floating and feeding on the food that is blown to them by the trade winds. Large strandings on Cape Agulhas suggest that numbers of hatchlings may be swept into the West Wind Drift south of Cape Town, whilst others may make their way into the South Atlantic in the stream of warm Agulhas Water which sweeps around the Cape during the South African autumn.

Loggerhead turtle laying her eggs on the Tongaland Coast of Natal. (George Hughes)

Loggerhead turtle returning to the sea after laying her eggs. Note research tag on her right flipper. (George Hughes)

For those hatchlings going into the West Wind Drift it is thought that their lives are cut off short as they might not be able to survive the very low water temperatures. This is, however, not certain as yet and there are indications that the young loggerhead is considerably more hardy than hitherto believed; recently a three-year-old loggerhead was found happily swimming around the ice floes in the Murmansk Sea just north of Russia. It was reported to be robust and full of life.

Added evidence to support the suggestion that the Tongaland loggerhead spends much of its time in the open sea comes from recent reports of thousands of young loggerheads passing the Azores and Canary islands at certain times of the year, and, after having their numbers thinned out somewhat by local curio dealers who turn them into tourist dollars, the 'fleet' then disappears for another year. As Western Europe is not, so far, known to harbour large colonies of nesting loggerheads, it is suspected that these wanderers originated on the east coast of the United States from where the Gulf Stream and North Atlantic Drift brought them across the Atlantic.

The southern Indian Ocean, has, unfortunately no islands such as the Azores, conveniently situated to harvest passing fleets of young loggerheads. But there have been reports of young loggerheads, approximately four years old by their description, arriving off Reunion Island at certain times of the year where their sojourns have been long enough for them to have earned the name of the blonde turtle (*La tortue blonde*). This is an excellent rendering of the loggerhead's colour at that age.

Rarer reports of even briefer visits have emanated from islands such as the St. Brandon group north of Mauritius, the Seychelles and the British Indian Ocean Territories, notably Aldabra Island. These more adventurous loggerheads are nearly always sub-adult.

Between five and six years of age the loggerhead once again makes appearances on the coasts of Africa and Madagascar. Whether it arrives there of its own volition or by a happy coincidence of the currents is unknown. It is still a surface feeder, consuming bluebottles, storm snails and being by now somewhat more active.

Over the next few years a gradual change comes over the loggerhead and it starts to explore the deeper reefs. Here it finds the molluscs, sea urchins and hermit crabs which soon constitute the bulk of its adult diet. The loggerhead has also been known to eat such indigestible items as plastic bags and bottles, whose presence in its gut usually has fatal effects.

It is suspected that the sub-adult loggerhead swims around the East African coast, exploring reefs and adjusting to the competition of other loggerheads until it finds an unoccupied feeding ground and settles down to regular existence.

During the five years of ocean drifting and the first two years of coastal exploration there is no external means by which the sex of the loggerhead can be determined. Preparation for adulthood commences around seven years of age. As far as the female is concerned this is of little bother externally so she does not change. For the male there are some quite startling modifications; the claws on the foreflipper start to grow and curl, forming a strong hook, while the tail begins to lengthen markedly, eventually reaching a length some five times that of the female. Both of these structural alterations have functional purposes, especially during the nesting seasons which lie ahead.

The exact factors that bring on the nesting urge are completely unknown. Nor is it known whether a neophyte nester in Tongaland was originally hatched in Tongaland. From studies carried out in other parts of the world it seems highly likely that the sea turtle returns to the beach on which it was hatched. Whatever the truth, there is no doubt that once a loggerhead female has nested in Tongaland she will return again and again with remarkable accuracy, sometimes coming ashore within a hundred yards of the last nest she laid there, two, three or more years before.

Whatever the stimulation, loggerheads start to arrive off the beaches of Tongaland during late October and early November, hundreds of turtles arriving at roughly the same time. Some are drawn from as far south as Durban, if not further, and as far north as Tanzania, 2 000 miles away.

Copulation has never been observed in Tongaland waters, but occasional males often become too amorous and chase females into the surf zone where they are inadvertently flicked on to their backs by the waves and have to be righted again by protection officers. The copulation act must be quite spectacular if a pair of turtles in the Durban Aquarium are indicative of normal behaviour. This pair, having had no contact with their wild relations for some nine years, suddenly took it upon themselves to mate at the same time that the wild loggerheads were gathering off the Tongaland beaches.

From this copulation and those of the green turtle observed elsewhere, it would appear that the female is the reluctant partner and males throw themselves on to the female's back in order to mount. This is where the male's large claws on the foreflipper come in useful. He uses them to obtain purchase on the slippery carapace of the female, eventually hooking the front edge of her shell. He then hangs on and brings his tail into action.

Had the male's tail remained the same length as that of the female, copulation would have had to take place face to face. The male, however, is able to curl his tail down under that of the female so that the cloaca comes into contact with that of the female. Once the cloacas are in contact the female grips the male's tail between her hind flippers, a double insurance against dislodgement and the possible loss of valuable fertilization. Copulation lasts for varying periods up to an hour in length. During this time the male does nothing but hang on and the female must do what little swimming may be necessary. Some couples, blissfully unaware of their drifting movements, sometimes find that they have drifted over a reef and the tide has left them high and dry. The more determined pairs ignore this inconvenience until copulation is complete.

It has been suggested in the past that a sea turtle female can store sperm for up to three years and use it as required. It has not been proven that she cannot do this, but it may seem unnecessary in the light of the fact that there are always males at the nesting beach during the season and fertilization could take place there and then. This is an aspect that will require further study.

Irrespective of when they are fertilized the female arrives off Tongaland with between 500 and 600 ripening eggs within her. After arrival, an initial batch of approximately 120 eggs are ripened, fertilized and shelled, at which point she overcomes all of her natural urges to stay away from land and makes for a part of the Tongaland beach. Whether she is guided on her first visit by some unconscious memory is unknown, but after darkness she cautiously emerges from the sea through the breaking surf and sits, possibly for the first time in ten or twelve years, once again on the lonely, open beaches of Tongaland.

After peering around for some time the female, satisfied that there is no immediate danger, crawls up the beach until well above the high water mark where she starts to move around in search of a suitable site. Often, for some unknown reason the turtle suddenly abandons the beach, sometimes having moved only a few yards since coming out, and swiftly returns to the sea, emerging again either later in the evening or on the following night.

Should she not abandon her expedition, however, after some initial

Green turtle flays the sand as she prepares to lay her eggs. (George Hughes)

searching, she starts to dig vigorously with her foreflippers throwing sand in all directions until she has excavated a body cavity in which she lies with her carapace level with the surrounding beach. Changing her tactics she then digs with the hind flippers, lifting out a cupful of sand at each scoop. When deep enough there is a slight pause as she straddles the hole and commences to lay her soft, round, white eggs in batches of one to five at a time. Approximately twelve minutes or so later all 120 eggs are laid and she gently drops sand into the egg cavity. When full she kneads the surface layers of sand and then disguises the nest site by throwing sand once again with her fore flippers. The whole operation can take as little as 40 minutes, after which the female returns to the sea.

During the next fifteen days another 120 eggs will ripen, be fertilized and shelled, after which the female will once again emerge to lay; and so on up to five times during a season.

The females, having completed their purpose, appear to depart immediately for other areas. One female was recently killed in Tanzania just sixty-six days after leaving Tongaland. She had travelled the distance at twenty-five miles per day, an impressive feat of endurance. It is deduced that she was probably returning to her feeding territory.

Although this particular female holds the long distance record for a Tongaland turtle, she cannot compare with the 3 500-mile swim of a leatherback turtle from its nesting beaches in Surinam in the Carribean to Ghana, on the West Coast of Africa.

Just how many times the loggerhead female makes her reproductive pilgrimage to Tongaland is not known. It is certainly not less than four times as this is the record number of times tagged females have come back to date. During future years it is expected that more will return, giving an even clearer picture of events. Strangely enough there does not seem to be any cyclic control of these arrivals because animals have been recovered after one-year intervals as well as two-, three-, four- and five-year intervals.

It may be that the older a female becomes the longer she takes to recover from the strain of a nesting season. No one is certain.

The answer to this question and to the many others that remain to be asked before we can fully understand the life history of the Tongaland loggerheads will, or rather, may come after many more years of research and conservation work.

Indubitably the protection of nesting beaches is essential because no matter what or who may be waiting there; protection officer or research worker, killer animal or man, a Tongaland female will return to her nesting beach. Her only urge is to lay her eggs so that her aeons-old line will continue from the mists of the past into the distant future.

Chapter 3

Madagascar's West-coast Islands

Names like Nossi Be, Nossi Iranja, Langi-Langi and Nossi Tani-Kely conjure up visions of remote islands and tropical hideaways. In fact, three of these names refer to some of the best tropical diving available in any ocean. And because these minor ports of call off the coast of Madagascar are so remote they promise a fascinating and almost virgin undersea domain. **Brian Rees** *details the attractions and the hazards.*

Our bubbles pinged sharply as they burst into the water at 100 feet, their erratic passage to the surface punctuated only by my rapid heartbeat spelling out Paul See's warning in peremptory tones: 'Watch out for the sharks at Sugar Loaf – we have had plenty of trouble from them there.'

The excitement of our first dive in Madagascan waters, heightened by the strange conical form of this rock rising sheer on all sides out of the sea had led us to disregard his warning completely; the vertical face of the old volcanic core had spawned only a small reef on one edge. We dropped away rapidly to a thermocline at about eighty feet, entering a realm of murky, colder water and I could sense that my buddy Lionel was not happy either. A sudden thump beyond our range of vision made me realize we would be no match for any large shark, should one appear. We were soon back in clear, warmer water and saw no sign of predators during our dive there.

The following day at Nossi Iranja (Isle of Turtles), was different. No sooner had I entered the water to fire at a huge kingfish which passed carefully within range, when a ten-foot Zambezi shark (*Carcharinus leucas*) came straight at me out of the turquoise-blue depths. The stump-nosed maneater brought with him an entourage of smaller, tropical black-tipped sharks to circle me on the bottom. The previous day's tranquillity had lulled us all into a feeling of false security and I had drifted about fifty yards from the comparative safety of the others near the boat. Luckily my spear had not penetrated its target and my calls brought the rest of the divers over. The incredible number and variety of large game fish provided their own opiate, for we soon forgot the constant presence of sharks. Working in pairs with one diver playing 'shot-gun', we

Tropical black-tipped shark in Indian Ocean waters. (Gerhard Lauckner: Seaphot)

easily picked off more than thirty game fish during the course of a two-hour dive.

The fast boat ride from our hotel on Nossi Be to Nossi Iranja had proved worthwhile. We returned again to explore its virgin reefs during the following weeks. Those of our party who did not dive were usually quite content to bump along in the boat, keeping a watch for rising bonito, or perhaps a basking shark on the surface of a placid, glassy sea. During our twenty days on the islands, we were constantly surprised by the calm windless mornings, for on the windward side of this great land mass off the East African Coast the monsoon blew relentlessly.

Apart from the diving, a visit to Madagascar is an experience on its own. The Malagasy people, pleasant, affable and always smiling, went out of their way to make us welcome. This reception became the norm during our entire stay on the island. Of mixed Asian, Arab and African heritage, the *indigenes*, tempered by French culture presented a happy relaxed geniality wherever we encountered them.

Our flight over the fifteen kilometres of tropical ocean between the mainland and Nossi Be had been breathtaking. Brilliant turquoise reefs beckoned on all sides. The islands themselves are interesting, stuck away on the west coast of the *Red Isle* in the Moçambique channel. Nossi Be's (the Big Island) support comes mainly from sugar cane, coffee and pepper, with an aromatic perfume base distilled from the flower of the Langi-Langi tree. Coconuts, mangoes and guavas grow everywhere – the fruit season lasts all year round and costs the visitor nothing. Nossi Be had much to offer inland as well; it is pockmarked by volcanic craters of all sizes. Numerous fishing villages cluster among the palms along the coast. The capital has the rather unusual name of Hellville; the town has a history chequered by privateers and mutinies dating back to the early 17th Century. Today, ten years after independence the area boasts a friendly dichotomy of French Colonial and Arabic architecture.

Our second expedition, together with families, was to explore the reefs and forest of Nossi Tani-Kely. The island is small – thus the name in Malagasy – perhaps 500 yards across and a few hundred feet high. On three sides the reefs drop sharply away to a sandy bottom at seventy-five feet.

An old French cannon removed from a wreck near Hellville, Nossi Be, Madagascar. It stands today in the main square of the town. (Gerald Cubbitt)

Beautifully-coloured fish
with a soft coral backdrop
are found everywhere along
the Madagascar coast.
(Peter Saw)

Underwater photography enthusiasts need go no farther; the island reefs are a sanctuary and no sea life may be removed or disturbed.

The corals in the area are exquisite, ranging from perfect specimens of fungi all the way through to elegant baskets and giant plate corals. These 'plates' attract hoards of reef fish of psychedelic variety. The island is also a regular haunt of giant game fish, marlin, tuna, tunny and bonito. The largest sea pike we encountered buzzed us constantly off the east point. On another occasion smaller fish milled round us in their thousands.

We found the water around the islands to be clear each day and the only ocean currents of any consequence arrived with the monthly spring tides. We were warned to keep clear of the western shore of Tani-Kely. Local reports indicated that the area was notorious for sharks, but there is something atavistic in the curiosity that draws the average diver to view a shark, and armed with a twelve-bore 'bang-stick' we were on the spot within days. For the purpose of attracting a shark we had purposely brought a small, freshly shot kingfish with us from another reef. This was hung below the boat to attract our quarry. For ten minutes or so we saw only a shy grey nurse, but with no formal introduction, a large Zambezi 'killer' appeared, first seen standing vertically in the water sniffing one of the divers' gun-floats. The newcomer – about twelve feet long – was obviously agitated. After some exciting minutes during which we all became confused as to who exactly was hunting who, we thought better of the exercise and returned to the boat. I saw two other large sharks that day off the island, neither with dishonourable intentions, but a fourteen-foot hammerhead was a little too big for comfort.

It was clear that the reefs in the vicinity of Nossi Tani-Kely had previously been inundated by shell collectors prior to restrictions being imposed. Enough good specimens remain to make finding one exciting, even though they must be left in their natural habitat. The Malagasy law with regard to the Marine Park is stringent; penalties are heavy.

On shore, the island provides a panoply of parodoxes. A single path leads up to the lighthouse keeper's house where he lives with a wife and thirteen children; even in such a paradise time obviously hangs heavily. Nearby a tall wild fig tree hung limp under the weight of large fruit bats which flew off shrieking when children roused them with stones.

Nearer the mainland, the much larger island of Nossi-Komba rises nearly 1 000 feet out of the sea. This is the 'Island of monkeys'. It is here that the last of the local wild lemurs live – strange mutations of the anthropoid family and found only Madagascar. Unlike the monkey, the lemur is a lethargic animal with staring, cat-like eyes. We visited the main fishing village on Nossi-Komba and were taken by the headman to the forests where the lemurs were waiting for their food. We had been warned to bring something along to rouse them. Ripe bananas and paw-paws brought them gangling down from above and the children fed them by hand. The variety we saw were black. The brown female is usually more shy, and invariably keeps out of range of cameras. The Madagascar lemur has a long fluffy tail; their little ones use this extension to hang themselves from their mother's waist with the last coil round their own neck!

We had little time to explore all diving possibilities off this island, but accounted for a few good kingfish at some pinnacles of rock in a channel a mile or two offshore. We did, however, find some unusual 'bowl' or independent coral here, which floats on the water when upturned. Bleached, the undersea growth makes an attractive fruit bowl.

The first week of our stay was spent systematically exploring most accessible reefs and islands in the vicinity of Nossi Be. These are too numerous to record in detail, each having its own character and variety of sea life.

Those reefs to the north of Nossi Be are generally shallow, easily accessible and prolific with marine life including crayfish. Most of the areas are sandy-bottomed. Overall the water is clear and we saw many turtles active in the acres of waving seagrass which stretches from shallow water to the deeper outer reef areas. In deeper waters we had a number of memorable dives out in the blue, chasing elusive bonito which would break surface in their hundreds while feeding on bait fish. They proved too fast for us in their own environment; the sight of them sounding a hundred feet below us was worth the experience.

Perhaps our most exciting dive away from the islands was at Five-Metre Bank, about five miles from the nearest island. Although the coral was disappointing, the terrain and fish life proved fascinating. We had studied the chart, and noted a channel – 150 feet deep – cutting through the bank. A scuba dive on the edge of this channel was as eerie as diving into a submerged grotto. The edge dropped away sharply and disappeared into an austere blackness below. I was almost gripped by vertigo on this drop-off; the feeling of weightlessness was strong. At times, I felt as though I could roll to the bottom of the smooth slope. Reef fish were abundant, and we saw a few large shoals of game fish patrolling the verge.

On returning to the reef with guns on a later expedition, we met three local spearfishermen patrolling this edge complete with aqualungs, but no large fish were bagged. They confirmed that the channel had proved a regular source of large game fish, but could not understand why we declined to spearfish with scuba gear.

An Indian Ocean flat rock cod (Dermatolepis striolatus). (Peter Saw)

They could also not understand our reluctance to shoot grouper or brindle bass. Judging from trophy pictures taken during previous visits, one would have expected that the brindle bass was the only fish in the sea off Nossi Be. We left them alone, for they rarely provide the same excitement and tenacity one experiences when tackling large game fish.

Later, when we returned with our first catches of kingfish, pompano and sea pike these local youths were astonished. They had never before seen these fish taken by spearfishermen in such numbers.

It was our turn for a surprise a few days later. One of the island fishermen returned to the hotel in his primitive dugout canoe – these boats go out with barely six inches of freeboard – with a huge ten-foot blue marlin onboard. He had caught the monster only a mile from the hotel. How he had subdued it working within the confines of such a frail craft would have provided a yarn in the tradition of Hemingway. We learnt later that he regularly brought in sail-fish from some unknown channel.

In contrast to the natives of Africa, the people of Madagascar have little inherent fear of the sea. We were constantly surprised to find youngsters diving from dug-out canoes with modern equipment, and shooting some big fish too. Malagasy divers also supplied the hotel with crayfish. One day they brought in fifty spiny Indian Ocean crayfish – no mean feat when diving from tiny craft with simple equipment. To them sharks were a hazard to be respected but not blindly feared; they saw them on just about every dive.

For us Hotel Coco Plage was casual and uncomplicated; an ideal diving resort. The cuisine was good. The chef excelled in the personal preparation of any guest's fish catch. The French *maître d'hôtel* tolerated our strange habits and odorous shells with a benign shrug of the shoulders. Nor did he mind our compressor. Each evening the machine chugged away, drawing clean air from the top of a palm tree to fill our aqualungs.

Saturday night was Casino night. Locals flocked to the hotel from plantations and villages for miles around to lose a few hundred cents at the wheel and dance in traditional style to the local band. The girls were fascinating. Most were dark skinned and danced in bare-midriff outfits with colourful ankle-length skirts. It was like something out of a Gauguin painting; they had their own particular grace, part Polynesian, part Occidental and chirped with delight as we tried to match the beat with them. Light and dark skinned, they mixed happily together. Their affluence and well-being indicated obscure but obvious connections with some of the wealthy White sugar farmers on the island.

Certainly the most bizarre experience in the area must be to attend a funeral procession. We met one mourning party making its way along the beach shortly after our arrival. The group carried a draped coffin on their shoulders. The heat of the afternoon and the soft sand made for an enervating experience. Every hundred yards or so the group would stop, open the coffin lid and remove a ten-litre can of palm wine. All the mourners would then sit on the coffin, and pass the tin around. We were offered some of the brew and not wishing to offend, brushed the dead bees aside to drink sparingly of a dark potent liquid which tasted almost like an Irish mead.

As divers, however, our thirst was for the sea. We never tired of its changing moods and colours. On a return trip to Nossi Iranja, we visited the southern reefs where we found the clearest water of any area dived in during our trip. Visibility extended over 100 feet in all directions. Prolific in both game and reef fish, we exhausted our bottles and film in a single dive in the area.

To do justice to the variety of these beautiful islands, any serious diver would need at least six months out of the cyclone season to explore the area adequately. Even then, many places would still remain little more than a name on a chart. We by-passed the huge 'Bay of Russians' to reach Sugar Loaf. On our short visit we were never able to visit the Mitsio islands far to the north of Nossi Be.

Paul See, a local resident and our attendant guide, offered sterling service from his Dive Shop and promised more facilities for visitors who were still to come including, eventually, a forty-foot launch complete with compressor.

With the vast Indian Ocean and Moçambique Channel as its backdrop and the islands providing the attraction, Madagascar's north-west coast provides much for the international diving community in the immediate future. Few who have been there do not promise to return.

Chapter 4

Encounter with a Stonefish

Regarded with awe by everyone who comes into contact with this fearful creature, the stonefish claims lives each year along the East African coast. The late **Professor J. L. B. Smith** *made a close study of this deadly fish during his Indian Ocean meanderings. His conclusions are fascinating.*

Probably the best known and most dreaded venomous fishes are those belonging to the group of the so-called scorpaenoid fishes. There are many different types, including the dangerous turkey-fishes and others like them, but the king of them all is the deadly stonefish, or rather stonefishes, for there are several species, all falling in the family *Synancejidae*. They are typically 'modern' fishes; that is, there are no known fossil types, and they are among the most highly specialized fishes in their particular group.

These awesome creatures are found in the tropical parts of the Indian and Pacific oceans, all living only in the sea. None is known to exist in fresh water. While differing in minor particulars, all the different species have the same fundamental nature, characters and habits; they are deadly brutes, greatly feared and hated by people of all races in the vast areas which these fishes have colonized.

Stonefish (J. L. B. Smith)

There are two different species in Australia, and another is known in India, while in East Africa there is only one *Synanceichthys verrucosus*, which occurs over almost the whole tropical Indo-Pacific.

All stonefishes are ugly, clumsy, almost evil-looking brutes that inspire horror and loathing at sight. They have a squat, lumpy body, covered with thick warty skin. The head is grotesque with large mouth and projecting lower jaw; the eyes are rather small, almost on stalks, concealed by flaps of skin and warts, and there are ridges and pits that make the whole head look like a rough piece of eroded coral or rock.

One of the chief characteristics of these stonefishes is that they are so shaped as to be virtually invisible when they lie among rough broken coral in a pool. They are unlike most other fishes in that they do not swim away when anyone approaches, but lie still, relying on their almost incredible home-built camouflage to render them invisible.

I have found them in pools, quite large specimens, that were almost impossible to distinguish as fishes even with one's eyes barely eighteen inches away. They lie absolutely still. Even their gills hardly move as they breathe, and it is not uncommon for shrimps to settle on them or for shellfish to crawl over them as if they were bits of old coral.

There is one species of stonefishes in the Pacific that closely resembles the dark volcanic rock on which it lives. Most others are grey or brown, exactly like their surroundings. The colour and

The Indian Ocean stonefish is grotesque in all respects and has claimed many lives. (Peter Saw)

markings vary. Some are reddish and most have spots or flecks of red, orange or yellow, much like rocks or coral. It is possible that the fish can vary its colour like a chameleon.

None of the stonefishes grow to any size. The largest I saw in East Africa was about fifteen inches long, the head five to six inches wide, the weight possibly two-and-a-half pounds.

Nothing was known about the early stages of the stonefish until we hunted them in East Africa and were able to describe the considerable changes that occur with growth. The young stonefish looks quite different from the adult. At about an inch in length, it has quite a normal shape, the head not markedly broad, but the body is covered with curious retort-shaped lumps that later develop into the skinny warts of the adult fish. With growth, the head rapidly broadens.

The habits of this repulsive creature are interesting. It is pre-

The stonefish is a rather ugly beast that has successfully colonized vast areas of tropical shores. (Peter Saw)

eminently a fish that exploits its environment to the full. Most live in the inter-tidal area, and those I have seen were in pools or in small channels where they lie concealed. They feed, apparently, entirely on fish. The small coral fishes obviously cannot see them as they lie on the bottom, and when any come near enough the stone-fish heaves itself up and gulps down one or more before they realize what has happened. Stonefish, apparently, never lack for food. I have found a number with the belly almost full of the small fishes common in tropical pools and channels.

It is clear that the stonefish emerges as a rather ugly beast that has successfully colonized vast areas of tropical shores. This alone would not make it so well known. Its real claim to notoriety is its venom; its deadly nature in this respect has caused it to be an object of fear and hatred to man from the earliest times. The Africans in all areas where it occurs know it well and teach the young its dangers.

The stonefish has thirteen dorsal fin spines on the back, each a highly developed venom apparatus. The whole fin is completely covered in thick warty skin; only the extreme tips of the spines are exposed and these are needle sharp and silky smooth with a deep groove on each side. About halfway down each spine there is a more or less kidney-shaped gland that holds a milky-coloured venom; each of these connected with the groove in the spine. The least pressure on the skin that covers the fin forces the venom from the glands through the grooves.

When any large animal approaches a stonefish, it does not move. All it does is to erect its spines. If a foot presses on them, one or more spines easily penetrate the flesh and the venom is injected under great pressure. With a moderate-sized stonefish, by pressing on this skin with a stick, I have seen the milky fluid shoot out to a distance of four or five feet, so great is the force.

This venom is one of the most virulent known. The first effect is to cause terrible pain that increases rapidly to such an unbearable stage that the victim becomes demented or delirious. If, as often happens, he is wading, he may, and usually does, fall into the water and without help could easily drown. In addition to the pain, the wounded area, and quite soon the rest of the limb as well, starts to swell and may attain great size. Apart from the killing pain, the venom has a serious effect on the heart, causing a dangerous or fatal drop in blood pressure, so that if it is injected into a vein, death usually results within one or two hours. It is as well to remember that the stonefish can live for several hours out of the water, for one that looks dead can suddenly jump and stab an incautious hand.

There can be few conditions as unpleasant, agonizing and distressing as that of a stonefish victim. Local anaesthetics and pain-relieving drugs have no effect on that pain. One medical expert has said that nerves can cause so much pain, and no more. He should try a stonefish, even a small one. The agony is unbelievable and rarely lasts less than six hours, but the wounded limb is painful for many days. The venom appears to kill the flesh around the site of the puncture, for this usually opens to a running sore.

Primitive people with no modern medical aid suffer in this respect as in tropical areas deadly infections are common, and any open sore can easily become infected, which often results in the limb being deformed or even lost. I have seen some awful cases in East Africa, twisted and deformed legs and feet; in one case there was little left but bone. One man showed me his twisted foot – he had been unable to walk for three or four months after being stabbed, and could only just hobble then. While some stabs result in death, even those victims that recover generally suffer from ill health for an extended period; some are maimed for life.

At Baixo Pinda in northern Moçambique, a part where we worked, boats laden with locals often go out to the reef on a falling tide. The boats ground, the Africans, all barefoot, usually wait until near low tide before going out along the reef to hunt fishes and other creatures for food.

On one occasion a man who got out first, when about waist deep not twenty yards away, trod on a stonefish. He yelled and went back to the boat, was in great agony even as he climbed in, fell writhing, and was unconscious in a few moments. The boat was hard aground about four miles from shore and before she floated with the rising tide, the man was dead.

We first went to Pinda, in Northern Moçambique, and worked for some months there. We were fully aware of the danger of the *Sherova*, as the natives name the stonefish. They have certain remedies, and believe in them, but we were not impressed and always wore thick-soled boots and rubber leggings. We encountered several *sherovas* which the natives always killed as they came across them.

Professor J. L. B. Smith and his wife Margaret working with specimens at Shimoni in Southern Kenya. Wasin Island can be seen across the channel. (Margaret Smith)

One afternoon, when we were working fairly close to the shore, poisoning a series of pools with the pounded bark of a forest tree (which brings the fish out from their hiding places), one of our workers shouted '*Sherova!*' and there was a lovely specimen, about seven inches long. Orders were always to put dangerous fishes straight away into a special bottle. One worker always carried on his head a metal tray covered with cloth for the larger fishes and for bottles and tubes. Having some rarities in a tube, and wishing to go to meet some canoes from the reef, I called him and got him to bend so that I could put the tube into the tray. As I did so, I felt a sharp stab and withdrew hastily, seeing two bleeding punctures in my thumb. As I looked, a stab of intense pain shot up my arm, and with foreboding I lifted the cloth. There, alive and with all spines erect was the stonefish, which this idiot, contrary to orders, had put into the tray.

Hastily winding some string round the thumb and cutting across the stabs, I told him to fetch my wife, who was working round the corner out of sight. Sucking the wound vigorously, I set off at full speed for the lighthouse, but before I reached the beach the pain was intense, and by the time I was climbing the old coral track, despite the tightly constricting string, it had become a searing agony spreading through my hand and arm.

That half-mile through the forest was a grim battle to retain consciousness. The road moved up and down, and the trees swayed wildly. I had to fight an insane desire to roll on the ground. It is probably this that causes victims on the reef to throw themselves into the water. (An Israeli scientist has recorded that when in agony from the stab of the related venomous turkeyfish he had the same desire.) At the lighthouse, I called the keeper to help me. We had one bottle of Nupercaine and – not more than half-conscious – with left hand and inexpert aid, I managed to sterilize a syringe to prepare an injection.

During my life I have endured a good deal of severe pain, but nothing comparable with this appalling agony. The sweat literally poured from my body and dripped to the floor. I could hardly see. My hand was swelling fast, and as I tried to fill the syringe, the precious bottle slipped and smashed on the dirty concrete floor. As I stood staring stupidly at this wasted fluid, my wife arrived and, having heard it was a *sherova* stab, soon injected novocain. After a few moments, the pain eased slightly, but remained almost intolerable, so that I remember little of what happened at that time. In that period, however, a native and his wife appeared, saying that they had a treatment that cured. My wife decided, in desperation, that they might try. The woman took my hand – that in itself an agony – and mumbled an incantation while chewing some green leaves. Then she spat on my thumb, rubbed the mess into the cut, called for what looked like hard yellow clay and rubbed it all over the thumb. Though only half-conscious, my mind said: 'Leprosy, syphilis, tetanus.' They said the pain would go in five or six hours after the treatment. Rewarded, they went away. We also went, but to use an antiseptic.

Half-fainting, with this incredible agony undiminished I staggered about, unable to keep still, until my wife gave me an injection of morphia. After some time this diminished the sweating and made me sleepy, but had no effect whatever on the pain. About three hours after the stab, my wife decided to try hot water, and within one minute of immersing my hand, the effect was dramatic. The agony was reduced to just a bad pain. Again fully conscious, I felt as if I had been dragged back from some terrible peril and suddenly found myself consumed by a raging thirst so that I drank innumerable cups of tea – the water there *must* be boiled anyway.

We continued the hot water treatment for some hours and the pain gradually eased. In the morning, glad to be still alive, I found my swollen hand and arm an awful sight. The thumb had gone black about the stab. Large painful blisters of yellow serum formed and, when punctured, dripped continuously for days.

The swelling of the hand and arm increased to a maximum after three days, extending even above the elbow. Five days after the stab, inflammation and pain increased alarmingly and pus appeared. This was treated with penicillin injections, which cleared the infection and halted further spread of the blisters.

Two weeks later, my arm was back to almost normal size, but could not be used. The hand and thumb were still swollen and painful. After thirty days, the black parts began to fall away leaving pink scars, but the site of the stab was still an open sore that did not heal finally until fifty days after the stab.

Even three months afterwards, the hand was still weak and the thumb barely movable and still tender. The hand remained weak for several years. The toxin has a most depressing effect and made old wounds and weak spots ache, much like an acute rheumatic attack.

We have that particular stonefish in formalin. He is a beautiful specimen of his kind, and we still look at him in utter disbelief that so much hellish agony lies locked up in those spines.

My records read:

> I have just been attending a funeral, a most unusual occupation for me, but this was a most unusual funeral, in an unusual locality. I am writing this facing the sea at the foot of the lofty annulated cylinder of the stately lighthouse of Pinda, in Northern Moçambique. This was once the

front room, but now the roof, the door and the windows are missing. All were ripped away in the cyclone of five months ago and with the clamorous calls from over a vast devastated area only absolutely essential roofs have been restored.

But to return to the funeral, held right at the sea: The corpse twenty-four hours ago was a healthy, tough Pinda African fisherman of between

The funeral of a stonefish victim, with the dead man's out rigger canoe in the foreground. The reef where he died is just visible on the horizon. (Margaret Smith)

thirty and forty years of age, an experienced seaman. Early yesterday morning, he and a companion, in their small dug-out canoe – fitted with outriggers, mast and reed mast sail – set off for the seaward edge of the reef, five miles out, well in time to reach their hunting ground before the falling tide exposed the raised rubbly edge. At spring tides, every man and woman who can get away goes out to that ten-mile-long ragged, narrow margin. Its teeming life is stripped daily and as often renewed when the more than twelve-foot tide sweeps over twice. Most go out in big sailing dhows that are grounded on the coral and later remain high and dry until the mill-race of the tide pours over the edge and refloats them.

From the sea in rare calm weather the whole edge of the reef looks as if it has been fortified with an irregular palisade, but the apparent stakes are humans: hundreds and hundreds of African men and women, probing the reef for its hidden wealth of food. Each one has a barbed spear and hardened pointed sticks and neither fish, seacat nor shellfish has much hope.

But it does not go all one way. The reef hits back. First of all, there are the wind and the sea. Many of the natives are shrewd judges of the weather, but they are fatalistically venturesome. Quite often a big blow sweeps up just after low tide. The larger boats generally manage to reach safety, but not always. About twenty years ago, the first keeper at this very Pinda lighthouse went out one morning in an eighteen-foot sailing dhow to the reef to fish. With him were his wife and two natives. In mid-morning a gale sprang up and not a vestige of those four humans was ever seen again, and in that same blow many smaller craft and their crew were lost as well. Such loss of life and boats occurs all the time.

Only last Friday, four canoes were overwhelmed, but the crews of three had '*boa sorte*' (good fortune) for they were seen and rescued by larger boats, but two men with their canoe did not get back. They were about five miles out. I saw something of that tragedy with poor field-glasses from the top of the lighthouse.

Then there are stingrays – hidden pain and sometimes death; stinging coral, which when touched, even lightly, brings searing agony that endures and leaves painful livid weals; the deadly *conus*, a small shellfish whose stab causes paralysis, ill health and even death, a constant danger to unwary probing fingers or to feet. Everything in this tropical sea seems to sting and inflict pain, but none is so hellish or deadly as the king of them all, the repulsive stonefish.

And that brings us back to the funeral. In their shallow canoe, our two men got nearer the reef than the dhows, parked their craft on a coral hump, and set off, wading thigh deep, each intent on his prey. Soon afterwards, the older man noticed that his companion was not visible and eventually seeing him in the water way off obviously collapsed, set off to his aid. He found him almost delirious with pain, but able to say that a *sherova* had stabbed him in the foot. The older man shouted to another African, who came to help. They marked the spot with a spear and carried the victim, by then delirious, to the nearest canoe.

The man who helped them went back to the spot and probing about soon found the stonefish under a nearby ledge. It was a big one and he killed it, then cut it open and took out the gall-bladder, which they believe is an antidote – a futile belief. When he got to the canoe, by then surrounded, the wounded man was unconscious and within a short time was dead. This happened not far from where my wife was working on the same stretch of reef. I was 'bombing' that day and did not see the corpse until later.

There was only one stab, in the second toe of the right foot, but probing showed it was deep, from the front right along the bone. As it killed this man so quickly, the toxin may well have been injected directly into a vein. It is as well to remember that Africans, as a rule, appear to feel pain less than we do.

So they took him to shore – two paddlers and the corpse in a larger canoe, while the news and the dreaded word '*sherova*' spread along the reef like a gust in ripening corn.

This man lived quite far inland, but because he was killed out at sea, custom demanded that he should be buried near the sea. Many of the people here remembered my own terrible experience at this very reef and lighthouse six years previously. They knew we were interested in anything to do with the *sherova*. They came to tell me the funeral would take place in the morning about ten of the clock – at the shore, where he was landed – only, please, not the *Senhora*; it would be strictly for males only. I found out that this held for all funerals, even when the corpse is a woman. 'Why?' There was shuffling and hesitation. 'Custom! But women cry too much.'

From his home, on his own stretcher bed, the victim was carried to the beach, then to the shade of a low tree on the very edge of the dune above the sea, shielded from the sun by the reed-mat sail of the canoe. As he died at the sea, he was put on the sail and washed with fresh sea water brought up in a special pot by the father. As they plainly found my presence embarrassing, I left them during that part. Then they tied him up carefully in his own blanket, with the hands folded modestly over the genitalia, the whole body stiff and stark; they kept it carefully shielded from the sun.

Much of this was done to a dreary eight- or nine-note dirge in a minor key that went on and on. They put him on his bed and formed a ragged procession that went along the shore, droning this mournful succession of notes, then up to the grave. A real seaman's grave it was, right on the margin of the beach, in sight and sound of the waves barely twenty yards away.

The digger was still hard at it when we got there. The bier was put at one end, while the experts debated depth and length. Suddenly there was an exclamation and, bit by bit, out came most of the bones of a human skeleton, plainly an adult male, clearly another victim of the sea, but unknown to the mourners. This disturbance of the remains caused almost unseemly mirth among the mourners, whose jokes I could not follow. Eventually, all were satisfied with the burial hole.

Those in charge, and the father, four in all, got into the grave. The corpse was handed to them and they placed it carefully on its right side facing the sea and against the seaward side of the grave. It was then fixed firmly in that position by twenty-five stout three-foot-long stakes forced in at an angle of about sixty degrees to the bottom, forming an almost continuous oblique palisade that virtually concealed the body. Bunches of coarse grass were then passed in and trampled firmly down by the four men in the grave, until the stakes were fully concealed. The four then clambered out. The exhumed bones were carefully thrown back, noticeably not touched by hand. Then the mourners crowded around the grave and started shovelling in the sand, to the same dreary mournful tune.

The filled grave was pressed well down, then they made the conventional hump and drove in four stakes to mark the place. So one more stonefish victim had gone to his last resting place; in his last moments conscious surely only of the sea and the pain.

After my own terrible experience, we were puzzled to find people (even medical doctors) on the coastal areas of East Africa, where stabs occurred, generally ridiculing the idea of danger to life from stonefish venom. Several told us they had seen cases, but beyond a little pain that passed in a short time, nothing further happened. Eventually, we found the explanation.

In East Africa, stonefishes are nowhere abundant. Indeed, in most populous areas, they are rare because the Africans kill every one they see, partly because they hate them and partly because they are good to eat.

But there is another fish, *Scorpaeonopsis*, an ugly spiny brute that looks very like the stonefish. We got to know him well and discovered that while his stabs cause considerable pain, this soon passes, and compared with the stonefish the effects of its stab are like a child's firecracker to an atomic blast. What the doctors described had almost certainly been *Scorpaeonopsis* stabs.

We also learnt later that quite often the venom apparatus of the stonefish can be partly defective and one or more of the spines may lack the venom glands. So there is a slight chance that a person could be stabbed by a spine lacking a gland and suffer little or no harm.

There was much confusion about the matter, and after I had published an account of my experiences following a stonefish stab, an overseas scientist wrote to say that he hoped I would not consider him heartless, but that after so much vague rumour he was really thankful at last to see a reliable report from a scientist!

While stonefish stab cases along East Africa have almost all been Africans, the reverse is true in Australia where they have stonefishes on their northern tropical shores. The species we find in East Africa, *Synanceichthys verrucosus*, does extend to Australia, but is rare there. They have another similar, *Synanceja trachynis*, that is nearly as bad, certainly as far as pain is concerned. Confirmed reports of death from those parts are rare and once again there were qualified disbelievers.

Despite the actual death of a man who trod on a stonefish there, as recently as 1956 a leading medical expert in Australia published a report 'de-bunking' the stonefish, stating that stabs are trivial injuries, and that deaths, if any, were caused by other factors. My report of the consequences of a stab from even a small fish and of actual fatalities in the Western Indian Ocean, changed the picture. However, others in the region of northern Australia had never agreed with the medico mentioned, and a full investigation of the venom, its nature, effects and counter measures was organized.

The average gland on a spine of the Australian fish was found to yield up to ten milligrams (one tiny drop) of the venom, which is a milky fluid. (My own rather casual experiments showed our stonefish to have very much more – several drops in fact from one spine.) The Australian scientists found that the venom is a solid held in suspension as a milky liquid. It is a protein and, like all proteins, is sensitive to heat. They were able to show that, when heated, it rapidly lost its potency – if kept at only 50 °C (120 °F) for a short time it became harmless – so the treatment of stabs by heat is scientifically sound. It destroys, or at least greatly reduces, the dangerous properties of the venom.

It was found (like many proteins) to be sensitive to changes in acidity, so that quite a mild acid or alkali also destroyed its venomous properties. It was also found to be destroyed or rendered harmless by potassium permanganate and by certain dyestuffs such as Congo Red. In experiments on mice and fowls, injection of the active venom caused much the same symptoms as in man, i.e. intense pain and a rapid drop in blood pressure. When injected into a vein, the fatal dose was very small. For a man, injected into the flesh, the fatal dose is estimated at fifteen to twenty milligrams; into a vein, far less, certainly well within the amount from only one spine of our African beast.

Much in the manner in which snake anti-venom serum is produced, Australian workers have managed to produce a similar preparation for stonefish venom. This helps to neutralize the dangerous fall in blood pressure. They also found that an injection of emetine hydrochloride solution, one grain per millilitre (which is acid in reaction) quickly relieves the pain. Even victims with several stab punctures who might well otherwise have died, soon lost all pain when emetine solution was injected into the stabs, and what is more did not suffer from gangrene and other secondary effects commonly following injection of the venom.

In the case of snakes, anti-venom of one species is not always generally effective, and it is not yet established that the Australian stonefish anti-venom will work as well for our beast. In any case, it is probably not available here. However, it seems likely that if emetine hydrochloride is injected soon after a stab, it will probably not only ease the pain, but destroy the venom as well.

Those in any danger of stonefishes should take ampoules of emetine hydrochloride solution with them (one grain per millilitre) and a syringe. Inject about one millilitre into each puncture, and it is certain that keeping the injured part in *hot* water will help as well. Not only does this apply to the stonefish, but it has been found that the venom of other dangerous fishes such as the sea-barbel and stingrays, so often the cause of great agony, is also sensitive to heat. It is therefore unnecessary for anyone to endure prolonged agony and evil consequences when stabbed by stingray or barbel. As soon as possible get the injured limb into water as hot as can be borne (do not scald). If hot water is not easily available, careful toasting over a small fire will do until hot water can be prepared.

If available, injection of the emetine solution can do no harm. There is every chance that it will work for barbel and stingray stabs. If it does, it will be a great boon to those who might otherwise endure not only great agony, but also, quite often, unpleasant consequences as well.

Chapter 5

Beyond the Turquoise Lake

A few hours' drive from South Africa's southern Indian Ocean shoreline lies a complex and intriguing maze of underground tunnels and caverns known as the Cango Caves. **Tom Hennessy** *dived and explored one of the adjoining undersea lakes – the Turquoise Lake – and his experiences were such that he never returned to dive there again.*

The water was as clear as Venetian crystal and seemed slightly greenish in hue, almost emerald-coloured in the light of the torch as the first eager eyes peered downwards at the mysterious new find. A roadworker's bulldozer had rolled aside a large boulder during the construction of a new route at Oudtshoorn near Southern Africa's Cape coast. Moving the rock revealed a yawning, cavernous hole dropping vertically into the unknown.

At first no one took much notice; for a while the contractors were conveniently able to dump all their loose rubble into the pothole instead of carting it miles to an assigned dumping area. But the capacity of the underground cavern seemed endless and refused to fill.

What did draw attention eventually was the fact that each time rubble was thrown into the hole loud splashes were heard from below. Finally someone decided to take a look and it was Mike Schulz, town clerk of Oudtshoorn who first mentioned something about an Emerald Lake. Having carefully secured a Jacob's ladder at the entrance to the mysterious hole, Schulz climbed down into the unknown to investigate. All he took with him was a hard hat and a torch. He recalled later that twenty-five feet below road level he found a subterranean lake about the size of a tennis court. He said that the water was so clear he could see passages leading off in various directions below the level of the water. Where did they lead to and how far did they extend?

It was these questions which finally brought a number of divers and spelaeologists on to the scene for it was immediately recognized that the Emerald Lake could provide the area – normally arid and continually pestered by droughts, although bordering on the Indian Ocean – with additional water. The South Africa Spelaeological Association was asked to nominate a diving group to investigate the cavern with a view to forming an estimate of the quantity of water available in the system. Thus the Iscor Underwater Exploration and Research Group carried out a preliminary survey in 1964 led by Laurie Short. They reported that a number of tunnels led off from the lake. Three of these were heavily blocked by vast amounts of rubble dumped there by the road contractors. At the request of Mike Schulz a second expedition was arranged by myself in 1966; he promised to lay on accommodation and power and lighting to the cavern.

A group of seven of us from Cape Town's Atlantic Underwater Club set off for the Cango Caves, armed with scuba gear, ropes, torches, compasses and considerable enthusiasm. We had precious little idea of the actual dangers associated with this sort of cavern diving, except what we had read of Cousteau's adventures in 'The Silent World'. Diveable potholes in the Western Cape are virtually non-existent, and most of our experience was limited to the sea, or zero-visibility mud dams.

The party consisted of seven experienced divers who were adaptable to most underwater situations. Apart from myself, there was 'Mac' MacLachlan, an expert on compressed-air machinery, with hundreds of dives to his credit; Nick Rutherford, today a Lieutenant in the R.N., then an engineering student at the University of Cape Town; Mike Hendra, a Cornishman, later to dive on the treasure flagship *Association*, off the Scillies; Werner Weiss, a German diver later to join Marine Diamond Corporation prospecting for undersea diamonds off South West Africa; Hein Blau, a veteran crayfish catcher and finally Peter van Eck, a founder member of the club.

We arrived at the entrance shaft to the Emerald Lake; it lay on the national road with a steep cliff on one side and a drop into a river on the other, A large makeshift manhole cover was removed and we gazed in awe as a Jacob's ladder snaked its way down the hole.

With considerable caution we descended to the lake seventy feet below, arriving there trembling from exertion (or nerves?). In fact this descent down the ladder proved to be perhaps the worst part of investigating the lake. The rope ladder twisted and bent over rocky projections placing a considerable strain on the descender, heavily laden with diving gear. Later we lowered the gear separately, taking little chance on possible damage to the apparatus. Powered by a 220-volt generator, a bank of bright lights filled the cathedral-like cavern with an eerie reflected glow.

The cave itself was quite large; its uneven roof sloping easily until it met the water level. From there on down the walls were perpendicular. In that light the sight was both intriguing and captivating.

Our first sortie disclosed four tunnels which led away below the surface of the water in various directions. Two led from the floor of the lake and were partly closed with rubble; two more led off horizontally from the precipitous sides and appeared to our uninitiated eyes as large forbidding holes leading into the unknown.

Immediately we noticed our first hazard: silt, an extremely fine dust accumulated over centuries. When churned up it reduced visibility to zero. We found later that when disturbed it took about eight hours to settle and proved to be a major obstacle in exploring the tunnel. Visibility was essential to estimate the quantity of water leading off the tunnels.

That evening we planned the next day's diving. It was decided to concentrate on the most navigable tunnel.

The initial plan was to dive in pairs to begin with, while visibility lasted. It was felt that two pairs of eyes on one clear dive would be more useful than two separate dives, where the second one would only be murky and visibility limited. The lead diver would carry a weighted line which would be dropped in the tunnel to serve as a guide for the next pair to follow. The divers would also be roped to the launching site in the cavern itself where someone would pay out a life line as well as a shot line, keeping a careful check of the length of rope paid out. Simple signals were used: one pull was 'are you O.K.'/'I am O.K.'; two pulls meant 'give me more rope'; three; 'pull me back'. The emergency signal was four pulls and the signal for a standby diver to enter the tunnel following the life line.

Since we expected the rope to pass around corners, it was the tender's job to be sensitive to all movement at the end of his line. Under the best of conditions this is no easy task.

In order to reduce the effects of mud kicked up by the divers near the launching spot, a large piece of hessian was laid underwater and weighted with rocks. This proved an excellent means of keeping the water clear and is to be recommended for future cave dives where silt is a problem.

The first diver to enter the water was Hein; Mac, his standby, checked his gear. Peter van Eck was rope tender and it was my job to control the shot line. Hein had merely to locate the entrance to the tunnel and drop the shot line at the opening; a simple solo dive to set the ball rolling.

Hein entered the water and sank out of sight his torch winking eerily as he gained depth and moved away. The steady burble of bubbles breaking the dark surface of the pool suddenly ceased as he passed under the roof of the cavern in his outward swim.

On his return Hein reported entering the tunnel mouth which started about ten feet under the surface and ran horizontally in a north-west direction. The opening was about six feet in diameter.

At the mouth of the tunnel, he said, another large opening appeared to drop away vertically but that was all he could see as his movement in the area had disturbed the silt and visibility had dropped alarmingly within less than a minute. In the confusion of being suddenly plunged into darkness Hein brought the shot-line back with him.

The start of our brief spell of cave diving had not gone well and because of the first mistake the rest of the divers that afternoon were led into a comedy of errors which would have done justice to a group of novices on a training session. Conditions were heightened by the fact that this was not only our first experience of diving in caves but also of caving generally.

Mac followed Hein into the water to find the tunnel but to lose the shot line and was obliged to return to base and try again. By now the water visibility had dropped alarmingly. He was followed by Werner, our experienced German diver who found a tunnel but concluded after a brief delay that it was the wrong one. On his second attempt he managed to find the correct opening and dropped the shot line just inside the entrance.

It was now the turn of veteran salvage diver Mike Hendra who, on entering the water found himself underweight and had to withdraw. Mac had another go, found the tunnel entrance and carried the shot line around the first bend in the tunnel but then promptly fouled his shot line with the pillar valve of his scuba set. He spent a determined ten minutes fiddling about underwater in a bid to unravel the mess while the rest of us waited anxiously on the surface. The only reassurance we had that all was well – though it was quite obvious to all of us that he had problems – was his peremptory answering jerks on the lifeline.

Having found more weights it was now Mike Hendra's turn to dive into the tunnel but he snagged the shot line in a crevice and returned.

We paused for a short rest. Tension had built up mainly because the Emerald Lake was a virtually unknown quantity. Someone jokingly referred to the presence of the green monster lurking at the end of the tunnel, and with this unlikely thought, Nick and I kitted up to dive along the tunnel. We were now diving as a pair. I was roped and Nick simply held my life line.

Our plan was to follow the shot line left by Mike and then move on. Within moments we had arrived at the entrance to the tunnel where a sharp turn awaited us. A moment later we hit a turbid patch and our torches were effectively extinguished. Suddenly I burst into clear water again, with Nick grimly hanging on behind me, not being able to see a thing. We arrived at the end of the line and moved slowly up the tunnel. At about this point we had paid out about forty-five feet of line.

Ahead, in the beam of our torches, the tunnel stretched out into the unknown, the water still as clear as we had hoped for. The limits of the tunnel seemed endless and this itself was a little disturbing as we were moving further into the cavernous depths. Behind us, billowing clouds of fine silt obscured the return path. Our bubbles rose and broke against the roof of the tunnel creating brief slivers of silver as they flowed to a higher point; below the floor was covered in mud a foot deep.

Slowly we moved on, playing our torches over the roof, watching for signs of projection. Momentarily a strange sensation came over me. I looked down and became aware of a huge hole in the tunnel floor stretching away before me into utter blackness. This vertical tunnel was also about six feet across and our torches failed to penetrate its depth. We hurriedly swam over the chasm. Ahead we could see large slabs of rock hanging from the roof; some had fallen on to the tunnel floor. Our feeling of uneasiness was heightened by the steady stream of falling debris from the roof. At one stage a loud clang startled us as a rock hit Nick's steel scuba tanks.

We paused to get a grip on ourselves deciding that it was too dangerous to proceed as a pair. I glanced to the right and discovered that another tunnel branched off the main one.

In moving over to inspect, we hit a patch of turbid water which had sneaked up from behind and suddenly all became black. I sank into the thick ooze, and lost all sense of direction, horizontal or otherwise. The shot line tangled around my right arm which carried the torch on a lanyard.

Grimly I held on to the life line in my left hand, for our very being depended on our not losing contact with the outside world, and tried vainly to remove the torch in an attempt to put some light on the tangle of rope. Meanwhile Nick hung on behind while we gyrated in the mud. At one stage I lost sense of the vertical, but recovered it rapidly as I sank back on to the mud. I was tempted to cut free the tangle, but dared not risk it for fear of cutting the life line. Getting out of the tunnel by feel alone would have been almost impossible because of the various intersections we had passed in getting that far.

There was only one thing for it; three pulls and get the hell out! This cave diving was not for me.

I felt for my life line which Nick was holding tight and tested the tension. It was slack! I slowly gathered in the rope and tugged, not knowing really in which direction to pull. Quite suddenly I was astonished to find that tension came from a totally unexpected direction. We were jerked back along the way we had come and most of the return trip was spent scraping along the roof. Unfortunately the shot line came back too. I only hoped that Nick had not lost his grip on the line as it jerked into motion. I was relieved to find his torch right behind moments after we hit clear water.

On our return, the tension at the base was terrific, especially for the next diver – Mike. We had only been gone about ten minutes, which seemed like an hour and had paid out about 120 feet of rope.

After disentangling the various lines the rest of us made a concentrated effort to relocate the tunnel in the murk, but failed. We then decided to quit diving for the day to give the mud a chance to settle. At our post-dive conference later we agreed that diving in pairs seemed a particularly clumsy method to explore the tunnel especially if one of the divers got fouled or entered a low-visibility patch. However, it was clear that two pairs of eyes on a clear dive was useful; after we had compared notes, Nick and I discovered that each had made different observations of the same item of interest.

The following day Werner was first to dive. He found the tunnel after one false start and dropped the shot line at its mouth, noting that visibility was still hazy.

I was next to dive in an attempt to regain the ground lost the previous day and swam rapidly to the end of the shot line, feeling a lot happier at not having the responsibility of a 'passenger'. On arrival at the farthest point we had managed to reach the day before, I looked carefully at the slate-like formations of rock in the roof and on the tunnel floor and decided it was perhaps a little too dangerous to proceed further down the tunnel. The danger of snagging the life line was real and I did not wish to take any further chances. After a ten-minute sortie I dropped the shot line, and signalled my return to base.

Nick followed the line out and spent six minutes inspecting the new branching cavern before quitting the main tunnel. He reported that the roof was crumbling in places. With this happy little homily, I persuaded Peter van Eck to have the last dive of the expedition. Until then he had been our rope tender and had provided excellent service. Mac took over tender while Peter cheerfully kitted up with considerable false heartiness. Our problems had not gone unnoticed to the man who had followed our progress with his fingertips.

Apart from a faulty regulator which was replaced, Van Eck's dive was uneventful although he did manage to venture deeper into the caves than any of us. Tension became unbearable as Peter slowly edged deeper into an unknown cavern. Foot after foot of line was played out, the littler marker on the rope bearing silent witness to Peter's nerve. We were aware, of course, that he could have been sitting comfortably just round a corner slowly winding in his own life line!

Mac anxiously signalled one pull. 'Am O.K.' Peter replied, and another three feet of rope moved out. We all imagined him to be suspended on the brink of an enormous bottomless cavern, similar to some of those we had seen in the nearby Cango Caves.

Mac sent out another signal. Unfortunately this pull turned out to be a rather powerful yank which jerked Peter back into his own cloud of mud. Another yank from Mac settled it; he gave up and signalled his return.

Unquestionably we were pleased to return to the sunlight although the trip had been exciting for all of us. What was clear was that further exploration of the tunnel complex would require more experience, more sophisticated gear and considerably improved techniques. A period of acclimatization, we found, was essential, and more time in between dives to allow the water to clear a little. Certainly diving in crystal-clear water down a deep fountain or large cavern such as the Rhodesian Sinoia Caves is a different proposition to diving along a muddy shaft hampered by minor rock falls and limited visibility.

We left behind one reminder of our presence in the Emerald Lake; our shot line. It awaits the next exploratory party who may follow it to the point where the unknown begins. For us our one experience of confined diving was enough.

Chapter 6

Mombasa's *Santa Antonio de Tanna*

In the shadow of Mombasa's 500-year-old Portuguese Fort Jesus lies the wreck of a ship which sank almost three centuries ago; the galliot Santa Antonio de Tanna. *Many valuable artifacts have been recovered from between the ancient craft's ribs and spars including Arab chain-mail, beautifully embossed bronze cannon as well as Chinese porcelain of the K'ang Hsi Dynasty. In the eyes of some of Kenya's veteran divers only the surface of this wreck has been scratched; much more still remains to be found.*

Diving into the murky polluted depths of any tropical harbour is hardly a cherished experience, especially when one is more accustomed to the clear blue depths of the South Atlantic and you have been told that only the week before a Japanese seaman was taken by a large shark about twenty minutes by boat from where you are about to enter the water.

So it was one cloudy morning in April, 1972 that we approached the site of the wreck of the Portuguese galliot *Santa Antonio de Tanna* which, the records tell us, sank in Mombasa harbour towards the end of the 17th Century. By the time we had donned suits and scuba gear we had one hour in which to complete our exploration of

Mombasa's old harbour with numerous Arab dhows in port. (Kamal Studio)

the site for the tide was about to turn. At that stage of the high water there would be little current on the site which lies within the ten-fathom mark.

Conway Plough was the first to go overboard followed in quick succession by David Brown and then myself. A few moments later we were heading downwards in a visibility which could not have been more than ten feet although I was assured it is as much as sixty feet at times. We were lucky, Conway said; visibility is often as little as five feet. I was also assured that there were no sharks. According to my colleagues, local African and Asian children regularly swim in the old harbour and to their knowledge no one has been taken by the predators on the Nyali Beach side of the island for years.

Entrance to Mombasa's 500-year-old Fort Jesus. (Al Venter)

Considering that we entered the water barely ten paces from the shore, the descent below Fort Jesus was surprisingly steep. The

harbour bottom fell away below us almost perpendicularly to black depths below and reached a maximum of about 200 feet towards the middle of the channel. Although I was aware that there were some of the finest coral beds within a few miles of where we were swimming the dark grey mud churned up twice daily by a fairly hefty current and speckled by the offal of ages, yielded little natural growth. Here and there a few clumps of sea grass had managed to take root. In between, an occasional rocky slab provided what sparse cover there was for a few fretful lobsters we passed on the way down. In places deep holes in the ocean floor about six or eight inches across presented the only other contrast, but I for one was not prepared at that stage to find out what they held; the East African coast has some of the most vicious moray eels I have yet encountered.

About halfway down I suddenly became aware of a massive shape, easily six feet across coming out of the murky depths towards us. Because I was separated from the other two by a couple of lengths they were not immediately aware of its presence. For a moment I froze, my breath coming in jerks through the regulator until I was able to make out the more wholesome lines of a large spotted *tewa* or grouper. It hung suspended for a few seconds its cavernous mouth gaping and then turned on its tail and was gone. According to Conway there are many of these brutes about but they never really troubled him.

At about fifty feet below the surface of the harbour Conway had already settled on a small ledge where some obvious mounds indicated the presence of the wreck. Little of the old galliot's

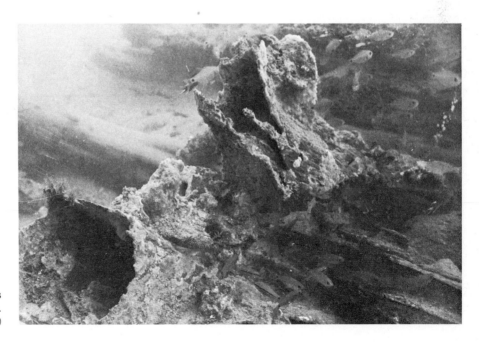

The sparse remains of this ancient wooden ship.
(Conway Plough)

original timbers remain although here and there a few rotten spars poke through. East Africa's tropical waters and marine borer worms have long ago reduced even these to a minimum.

The remains of the old ship as we settled on her that morning indicated that she has a breadth of about twenty feet. On the top side of the ship's list the depth at high water springs is thirty-five feet and the low side between forty-five feet and fifty feet. The galliot's length is roughly 130 feet but it is not possible yet to determine which end denotes stern or bow. In an attempt to plot exactly what lay where, some effort has been made to lay out a grid although this had not been maintained. What has been kept is a careful record of all relics brought up from the *Santa* and where exactly they were found. And what a fascinating array these are.

Among the items brought to the surface from between the crusty old ribs were pieces of chain mail which have been identified as either Persian or Arabic and have been dated at more than 300 years old; a beautiful Portuguese bronze breach cannon complete

Bringing ashore a coral encrusted cannon from the wreck of the *Santa Antonio de Tanna*. (Conway Plough)

with a fine inscribed boss; a twenty-inch high copper cauldron and a variety of wine and grain jars most of which were handed over to Mr. Jim Kirkman, curator of the Fort Jesus museum. In Mr. Kirkman's eyes these items make the wreck the most important historical find since the discovery of Gedi, the ancient Arab coral city in the forest near Watamu. Over the centuries Gedi had been

A copper-bronze Portuguese breach-cannon dated 1673 brought up from the wreck of *Santa Antonio de Tanna.* (Al Venter)

Conway Plough surfaces with two artifacts from the wreck of the *Santa Antonio de Tanna.* In his left hand he holds a Portuguese wine jar—in his right a K'ang Hsi Dynasty porcelain plate. (Al Venter)

pillaged and most of its treasures removed. The galliot, said Kirkman, who has spent most of his life doing archeological research in East Africa, had been hidden by the sea and it took 300 years for someone to find it.

Other, more valuable items brought up from the *Santa* during the latter part of the excavations included six porcelain Chinese plates twelve inches in diameter, *Famille Verte* of the K'ang Hsi Dynasty; all are fungus marked and four are complete. There was also recovered a tiny glass vial about three-and-a-half inches high; identified as the upper portion of an hour glass.

It was Conway Plough and his co-diver friend Peter Phillips who in 1966 originally stumbled on to the idea that there may be a wreck lying within a stone's throw of the ramparts of the 500-year-old Portuguese fort. Several artifact finds had been made in the area and the two amateur underwater enthusiasts guessed these were the remains of a ship. Together the two men, helped by Conway's wife Valerie, an accomplished diver in her own right, spent some time going over the steep terrain. Finally they found the narrow shelf and after quartering the area they examined a number of peculiar coral growths under which they found traces of timber.

This 20 inch copper crucible was taken off the wreck of the *Santa Antonio de Tanna*. (Al Venter)

Most of what has been brought up from the wreck to date was recovered from the ledge outside the outline of the ship, as though they were in use at the time and were washed off the deck when the ship sank. The copper cauldron, for instance, was found upside down and stuck in the mud. It could have been a rock except, when on closer examination, Conway Plough could see what appeared to be a row of large studs. When he tipped the bucket upside down he saw two handles. It too has been polished and cleaned and is on display in Fort Jesus.

For some time there was confusion as to the exact identity of the ship which lay for so many years at the bottom of Mombasa's old harbour.

As is recorded on October 20, 1697, the *Santa Antonio de Tanna* was moored in deep water off Fort Jesus when the cables parted and the ship was carried across the harbour on to the reef north-east of the island. Having cleared herself with the rising tide, she was eventually towed close to the fortress where it must have been anchored, but was lost shortly afterwards.

Mombasa's old Arab harbour. The wreck of the *Santa Antonio de Tanna* lies just beyond the flag to the right. (Al Venter)

The assumption is that having already lost one if not two anchors when originally moored, together with a considerable amount of cable, the vessel probably used a make-shift anchor and a heavy rope hawser, and was anchored close by the Fort on the edge of the reef. The edge of the reef slopes almost precipitously down into the deep channel of the old harbour, and dries out at low water. At this time of year the wind tends to go round to the east and sometimes increases in strength slightly round about dusk.

The possibility is that the vessel was laid to her anchor close to the edge of the reef on the spring ebb, i.e. with her bow pointing up harbour; high water springs are approximately 17.00 and 05.00 hrs. Probably about dusk the wind got up slightly from the east and swung the vessel over the shallow edge of the reef, where she grounded. As the tide ebbed further the vessel would probably have taken a dangerous list to starboard down the steep edge of the reef, and towards low water may have capsized to starboard and filled. In so doing the hull, weighted down by ballast, would have slid slowly down the edge of the bank; but the buoyancy of the masts and furled sails filled with air would have tended to right the vessel, and one would imagine she would have come to rest on the bottom of the harbour in a fairly upright position, facing up-harbour and close to the edge of the steep bank. It was in this position that she was found by Plough and Phillips.

More light on the subject has been extracted from Lisbon sources by Captain António Marques Esparteiro:

Fort Jesus had been besieged by the Muscat Arabs since 1696. The 40-gun frigate, *Santa Antonio de Tanna*, built in Portuguese India in 1681, left Goa for Mombasa with aid for the Fort on 15 November 1696, in the squadron of Admiral Luis de Mello de Sampaio, with an armament of 50 guns. Having left supplies and carried out a mission to Moçambique, it returned to Mombasa on 16 September 1697, with further aid.

It was moored near the outwork of the fortress with anchors fore and aft, so as not to swing with the tides. Under heavy fire from the enemy batteries, it was set on fire by the chance explosion of a barrel of grenades, but fortunately this was extinguished by the crew, though an attempt to change the moorings was unsuccessful, because the anchors were held fast in rocks on the bottom.

On 20 October the cables broke and the ship, with 150 men on board, was carried on to a reef north-east of the island, where it lost its rudder. When the tide rose it cleared itself from the reef and was swept near to an enemy battery, to the great danger of the crew. A party of 28 men from the Fort succeeded in taking the battery and towing the frigate close to the fortress, but the frigate was lost, how it is not exactly known, shortly after 20 October 1697.

It is interesting that almost all fifty cannon on board the *Santa* when she went down still lie at the bottom of the sea. When finally recovered they will certainly make a rich haul for all are certain to be bronze and embossed with traditional Portuguese markings.

But even after the wreck had been discovered and excavations underwater were continuing there were problems. Shortly after Plough and Phillips brought the first artifacts to the surface an amateur archaeological squad composed mainly of casual divers resident in Mombasa was formed to work on the wreck. With the help of a number of local firms a lifeboat and a compressor which operated an air lift in the form of a six-inch suction hose was provided. This was used to pump out mud from inside and around the wreck.

Work had hardly begun when the operations boat and all its gear was sunk, almost certainly by a horde of swimmers who clambered aboard to use it as a diving platform as soon as the divers departed.

Although the boat was soon refloated and all the gear recovered, the mishap did cause delays. This problem was to some extent offset when in November and December, 1970, two American divers Dennis Burkhart and Kevin Scherzinger were loaned by the United States Peace Corps to help work the wreck with local enthusiasts. In the course of these operations another large jar, more cannon, a Portuguese fire pot grenade, a Chinese jar and a rare salt-glaze beaker were recovered.

The work already carried out on the site has yielded items of considerable interest. Throughout the excavations, however, the resources of the small band of amateur divers who can only spare time after work and over weekends have been inadequate. A much larger compressor and a lighter airlift are necessary if the fullest use is to be made of their time.

There are plans in hand for raising more money to work on the wreck, and these have been given considerable impetus by the comments of Conway Plough in 1972. It is his view that the *Santa Antonio de Tanna* is only one of two or three ancient wrecks lying in the mud at the bottom of Mombasa's old harbour. Traces of other wrecks are evident, he maintains. All he needs now is the time to find them.

Chapter 7

In the Wake of the Coelacanth

The discovery of a live Coelacanth almost 80 million years after it was believed to have become extinct is well documented. For many years Professor J. L. B. Smith – to whom this book is dedicated – searched for this elusive four-legged creature. **Margaret Smith,** *wife of this famous Indian Ocean pioneer and today Director of the J. L. B. Smith Institute of Ichthyology tells of their earlier ramblings throughout an ocean about which less is known than any other major water-mass in the world.*

It was dark – dark and lonely. Around me black inky waves lapped and spat at the small motor boat as it chugged along at a maximum speed of four knots. The craft was a tiny speck in a black universe; the sea stretched away into the darkness. Above me were a few feeble stars gradually being blotted out by clouds as they drifted across the vast expanse of the Indian Ocean ahead of us. I was very conscious of the deep mile of water below me while the cockleshell of a boat was being buffeted and rocked by the waves and wind.

'Keep her pointed towards that star,' J.L.B. had said when he handed over the watch and the helm to me, 'and keep a look out for the light of the Pemba lighthouse.' He and the two Kenyan Africans then settled down to sleep.

Ominously rising cloud soon blotted out 'that star' and I frantically chose another, and yet another, all the while straining my eyes to see the light from the lighthouse. Sparks appeared in various places where I thought the horizon should be; the darkness and utter loneliness seemed to enter the very marrow of my bones. What madness was this? I thought with longing of my bed back in Shimoni and at about 1.30 a.m. I saw another spark, strained every sense, then relaxed a little, thinking it was as usual just another trick of the eyes. Then suddenly out of the darkness a glowing orange-red ball of light seemed to roll from the horizon straight across the sea at me. I ducked instinctively as it reached the boat and it rolled back again to disappear over the horizon. This time I knew it was no trick – it was the lighthouse that had sent that ominous red ball over the surface. My course was correct – that was a relief – but why were the flashes red? Red indicated danger and at sea this generally meant reefs. Then I saw that the flashes, twenty seconds apart, were alternately white and red, and I felt reassured. Alone with three sleeping men on an uneasy ocean, I watched for each flash. It seemed to falter and once I thought it had stopped, but reluctantly it threw its beam slowly and deliberately over the sea. Each time it lingered a moment on our boat, and then passed on, becoming a small speck of light and then blackness. But the light was an ally; I was on course. So I relaxed and not for the first time did I marvel at what a woman will do for her mate.

Here was I, originally a complete landlubber, born and bred in the shadow of the Stormberg-Drakensberg mountains, married to my former lecturer in organic chemistry, James Leonard Brierley Smith, who was nineteen years my senior. My twelve-year-old son, thousands of miles away to the south in South Africa, could so easily be orphaned while we worked along the dangerous and inhospitable tropical coast of East Africa.

Ichthyology seems a long way from organic chemistry. You may well ask how the brilliant young man who in 1916 passed first in Chemistry for the Bachelor's Degree in the Union of South Africa came in 1952 to be collecting fishes in Zanzibar waters.

Many years before, this same man, then a small boy, pulled out a shiny silvery fish from the Knysna Lagoon and the magic of that moment eventually changed his whole life. With his health impaired by the 1915 East African campaign against Imperial Germany he was finally advised to give up the long hours in laboratories spent on chemical research. But being a research worker by nature, he turned to the fishes he caught, and in 1931 published his first modest ichthyological paper.

For nine years J.L.B. had been publishing papers in both ichthyology and chemistry. He paid special attention to the sea fishes from Knysna to the Umtata River, an almost unknown area ichthyologically-speaking, and by 1938 he was honorary Curator of Fishes for the four museums in Port Elizabeth, East London, Kingwilliamstown and Grahamstown.

We had been married nearly nine months when, in January, 1939, he received a letter that was to colour the rest of our lives. It was from Miss Courtenay-Latimer of the East London Museum, and included a sketch that he immediately recognized as a fish believed extinct for millions of years. Later, in East London, he confirmed this and his announcement of the discovery of the five-foot living coelacanth, a close relative of the ancestor of all land vertebrates (*Eusthenopteron*) that had left its watery domain over 300 million years ago, rocked the zoological world. He produced a lavishly illustrated monograph on the mounted fish, and then together we set out to find the home of these primitive fishes.

However, the war intervened, but by the end of it we knew that the East London coelacanth was a stray. J.L.B. deduced that it had been brought down the east coast by the Moçambique-Agulhas current from areas too rocky to trawl, and remote from scientists and collectors. Unable to finance an expedition to seek one, in 1947 we had leaflets printed showing a photograph of the East London coelacanth and offering a reward of £100 (R200) for a specimen. These were sent up the east coast of Africa and to Madagascar, and we hoped that our propaganda would set thousands of eyes looking out for another of this rare species of coelacanth. In the meantime two other great events had shaped our lives.

In 1946 the newly formed South African Council for Scientific and Industrial Research enabled J.L.B. to devote all his time to fishes, and Rhodes University established the Department of Ichthyology

The Smiths examine the head of a giant parrot fish caught off Shimoni during one of their early expeditions to Zanzibar and Kenya. (Margaret Smith)

for his work. We had a part-time secretary, a messenger-cleaner and myself as his unpaid assistant. Giving up chemistry was a wrench, but both chemistry and ichthyology were expanding so rapidly that he had at that stage to make a choice. He was the only fully active ichthyologist at the time in Southern Africa.

Shortly afterward, a group of businessmen approached him to write a book on South African fishes. He refused to consider the project unless every available fish could be illustrated. I guaranteed to find artists from the University's School of Art and the *magnum opus* was started. His knowledge, experience and fantastic memory enabled him to write the book in three years. It was an experiment which gave the layman the power to identify most of the 1 250 species of marine fishes then known from Walvis Bay to Beira. I found four artists (three students and a young married woman) who managed to fulfil the requirements for accurate scientific illustration. When examinations and family commitments intervened I stepped into the artistic gap myself! From that day on I drew, photographed and painted fishes. Instead of being a musician, for which I had trained, I became a fish artist. Instead of working at physics and chemistry, my whole life revolved around ichthyology.

The Sea Fishes of Southern Africa (Smith, J. L. B.: Central News Agency; South Africa, 1949, '50, '53, '61, '65, '70 and '72) was published in June, 1949, and the first edition of 5 000 copies was sold out in three weeks. Long queues formed outside bookshops and the publication made a tremendous impact on the outside world. It is still used wherever marine fishes are studied.

We were now free to search for the coelacanth farther north. The western Indian Ocean, certainly the east coast of Africa, was almost virgin territory for any marine scientist. A German, Professor Peters, working on Zambezi fishes, had collected a few marine fishes along the Moçambique coast, and a Britisher, Lieutenant-Colonel Playfair, had collected fishes at Zanzibar during his office as Her Majesty's political agent and British Consul from 1862 to 1866. But that was all.

It was partly the search for the coelacanth and partly our own scientific curiosity that pushed us farther and farther up the East African coast. We journeyed north along the Moçambique coast, round the bulge to Mocambique Island and reached the barrier reef of Pinda.

The year was nineteen hundred and fifty. The dhow keeled over and I held my breath as we skimmed over the turquoise sea, the sails seeming almost to touch the water. My twenty-five dusky companions, however, chattered away unconcernedly and I realized that there was no real danger. The monsoon was fierce but we were partially protected and it carried us swiftly to Pinda Reef.

Ancient cannons on Pinda Reef, Moçambique. They are probably Portuguese and date from the 16th or 17th centuries. (Margaret Smith)

Most of the passengers were women of the Makua tribe, sailing out to the reef to collect sea creatures for food as the water receded. The crew consisted of four or five Makua men, their bronze muscles rippling in the sunlight as they handled the dhow, and the remainder were my own men, mostly Makua sailors seconded to us by the Portuguese Government.

Abdullah Bwana, an accomplished seaman, black as the proverbial ace of spaces and sixteen years my senior was my faithful, intelligent guard and chief assistant. His checked linen cloth (usually worn as a skirt over official-looking shorts) was twisted round his head in a massive turban. Standing on the prow and holding the mast with his left hand, he shaded his eyes with his right and peered ahead into the riot of greens and blues of the water to pick a course between the coral heads. I was transported into another world. Here was I, part of an adventure, a real life adventure, sometimes dangerous, always exciting, just what I had dreamed of. The seamanship of these black people was strange to a South African accustomed to the Xhosa dislike and distrust of the sea. Arab influence, and maybe Arab ancestry, was responsible for some of these men thinking nothing of sailing a dhow more than two thousand miles across the ocean to India. Abdullah himself was proudly Islamic – I learnt much of their customs and beliefs from him, and although he could not write Portuguese, he wrote Arabic with facility.

The sturdy Pinda lighthouse was some five miles behind us when the dhow finally grated to a stop on the inner lip of the barrier reef. I soon learnt to carry canoes with us so that immediately the dhow grounded we could paddle to the exposed parts of the reef and start our collection on the receding tide. Only Joachim, the policeman or '*Sepoy*' attached to our party, did not enjoy these trips. He was definitely not a seafaring man, but belonged to a hunting tribe. Perched in front of me as I paddled the small outrigger canoe, he would eye the waves fearfully as one or two slopped into the canoe. I could not resist asking him if he enjoyed it. 'Yes, *senhora*,' he answered in a quavery voice, 'because it is my duty!'

Joachim was worth his weight in gold when we were reef collecting. He was the perfect alarm system. To be caught by the tide in that area where it rose nearly sixteen feet during spring tides could be fatal. However, I had no need to watch the water. Joachim did that and long before the water lapped the reef top and came tumbling over in a rushing river, he was at my side urging me to embark.

I collected mainly by using rotenone, a substance that constricts the blood vessels in the gills causing the fish to emerge from the coral or sand. Prior to learning to dive, my collecting was restricted to pools and shallow water. Afterwards skindiving enabled me to work around coral heads in deeper water.

Before the use of free diving made it possible for ichthyologists to use fish poison at depths of 100 feet or more, the only way that fishes could be collected from the deeper water, other than by hooks, traps and trawling, was by the use of explosives. It was a collecting method used by all the important expeditions working among coral. We were no exceptions.

While I used poison, J.L.B. went to work with explosives, but fishes without airbladders are not affected by the latter. He would examine the coral formations through a glass-bottomed bucket, decide on the size of the charge, estimate the length of fuse, then carefully lower the 'bomb' so that the current took it to the required position. His experience and care enabled him to be remarkably successful in placing the blast so that fishes were brought to the surface. We always had to work quickly, netting them before they recovered or were eaten by birds or sharks. Contrary to what one would expect, sharks are attracted by explosions, and, having no airbladder, are unaffected by them.

Bombing was always exciting. We never knew what was going to rise to the surface. Every day we discovered something new to science or to Africa. In the isolated areas where we worked, it was possible to use explosives with the minimum damage. Although we had the necessary permits to carry and use explosives, we always collected as far away from other humans as we possibly could. J.L.B. was careful to cause the least possible damage to the coral reefs. In remote areas like the Seychelles, explosives still have to be used by the authorities to keep the channels through the fringing reefs navigable. In Mauritius, however, I saw dreadful damage caused to the coral reefs and coral heads by poachers.

I remember one day off Rongwi Island in Northern Moçambique I was rowing the canvas dinghy while J.L.B. put down a charge. Within a short time there were a number of sharks circling us, snapping up the stunned fish. This was the nearest I had even been to a large hammerhead shark, and I did not appreciate the way he looked at us. The water had fallen too far for us to return to the *vedeta*, a Portuguese cabin-cruiser type boat, inside the reef; so we had to bide our time hoping that the sharks would be satisfied with the fish.

At Pinda, maneating lions on shore, sharks, stonefishes and high winds at sea added spice and danger to our collecting. What excitement each new day provided.

Fishes new to science or new to the western Indian Ocean, some known only from the Pacific, appeared in every collection we made. What a thrill it was to stand on a reef or dive among the coral gardens, knowing that we were the first scientists to study these reefs. Who worried about heat, wild animals or dust while driving along jungle roads to reach more bays, swamps or reefs where yet more fishes waited to be discovered?

In 1951 we returned to the Island of Moçambique and Pinda for a short visit, then continued our investigations up the coast. At Port Amelia we were given a *vedeta*, the *Porto Amelia*, a slim forty-foot diesel-engined cabin-cruiser. She took us to Ibo, to the Quirimba Islands that lie north of Port Amelia and finally to Palma, the northernmost port of Moçambique. It has always been my dream that the beautiful coral gardens growing in the shelter of these islands north of the old historic harbour of Ibo will one day be proclaimed protected underwater marine parks. As a tourist attraction they almost equal the Great Barrier Reef of Australia, especially if the African hinterland can retain some of its wild life, providing the visitor with a double attraction. But they must be protected or they will be lost forever. The delicate ecosystem is so easily damaged as coral reefs are the life centres of the clear tropical seas.

The exposed reef provides a fascinating exploration area. (Margaret Smith)

We caught numerous game fishes on lures as we journeyed north to Palma. From Palma to Cape Delgado we travelled overland, at one point driving straight through the flames of a raging bush fire, with explosives and detonators on the back of our lorry.

South of Cape Delgado the currents were strong and treacherous, and the southerly monsoon made our return journey a nightmare. Our crew of black sailors marvelled at the power our charts gave us, and one day, when we ran for shelter behind a headland, I actually used some of the mathematics I had learned at school to draw a sine graph without tables to help me. The skipper had said that there would not be sufficient water to float the *vedeta* at low tide. So with my graph and his tide tables J.L.B. estimated that we would have nine inches of water below the keel at dead low tide. He was proved correct.

It had been fascinating voyaging between the islands. Not very comfortable in windy weather, but a sheer delight on calm days, the darker blue water marking the channels in the kaleidoscope of turquoises, greens and blues we sailed through. Far more comfortable, I reflected, than the inky voyage from Shimoni to Pemba.

It was in Kenya that I first learnt to dive. I had seen my first mask at Pinda the year before and realized what an advantage skindiving would be to collecting. But first I had to overcome my husband's opposition. It was his opinion that diving was too dangerous for a woman. Moreover he had no desire to lose his scientific assistant, his artist and, incidentally, his wife. It took all the feminine tact I could muster to persuade him, but finally at Malindi I slipped into the water equipped with face mask, snorkel and flippers.

It was wonderful. From the start a new dimension had been added to my life. The lovely blue of the water merged somewhere in the distance with a murky blackness where (was it my imagination?) large shapes seemed to be moving. On the sides of the reef were huge clusters of branching coral, solid coral in massive imposing heads, dainty bouquets of flower coral and brain coral in a myriad of vivid greens. Every colour of the rainbow was there, washed by the clear depths of a tropical blue ocean. Occasionally a large fish materialized. I held my breath as they came near, but they were invariably suspicious and veered away.

Carberry, my diving companion at the time - known as J.C. to his friends - went into action. Diving silently, he stalked a large fish, then fired his spear. There was a sudden flurry as J.C. rose to the surface with a struggling kingfish at the end of his line. Then it happened - all within a matter of seconds. From the murky depths below, a huge shape shot up towards the struggling fish. Moments later a swift torpedo-shaped barracuda came into view and without preliminaries its jaws snapped the fish in half. J.C. and I did not wait; we streaked for the boat, although even then I could not take my eyes from the barracuda as it sank slowly downwards with the remains of its victim.

From Malindi we finally moved down south to Shimoni where the coral round the offshore islands was wonderful and provided us with a wealth of material.

Netting for fish off Chwaka, Zanzibar. (Margaret Smith)

It was on September 23, 1952, in Zanzibar that I met Captain Eric Hunt, a man who a short time later was to have a marked effect on all our lives. At an exhibition we had arrayed interesting fishes caught in Zanzibar waters. Hunt came across the reward leaflet that we had had printed offering £100 (R200) for a coelacanth. He asked if the fish was likely to be found around the Comoro Islands, since he traded between Zanzibar and the Comores. Excitedly I

The town of Domini, Anjouan Island, Comores Archipelago is a favourite though rarely visited spot for European divers. The Comores waters are regarded as among the clearest in the world. (Gerald Cubbitt)

answered: 'Yes – very likely.' It was the one place that we felt the reward leaflets had not reached. The islands lay in the south-flowing Moçambique Current and were remote from museums and scientists capable of recognizing a coelacanth. Also, the steep rocky slopes of the islands would provide likely shelter for the rock-haunting fish the coelacanth's heavy armour and limb-like fins indicated.

Thus it came about that about three months later, on December 20, 1952, the second coelacanth was saved for science, and the home of the coelacanths – the Comoro Islands – was found. A five-foot coelacanth was about to be cut up and sold on the Mutsamudu market on Anjouan Island when it was recognized as the same fish as the photograph on the reward leaflet. By a remarkable coincidence,

Moments after the positive identification of the second coelacanth in French Comores. In the front row, extreme left is Eric Hunt, the man who made the discovery. Dr. J. L. B. Smith has his hand on the head of the fish while the French Governor of the Comores is holding the dorsal fins. (Margaret Smith)

at that very time Hunt's schooner was anchored off this island. The coelacanth was taken to him and he immediately preserved it with salt, rushed it to Pamanzi Island where he obtained formalin from a medical doctor, and sent a cable to us in South Africa.

In his 'scientific thriller', *Old Fourlegs*, Professor J. L. B. Smith tells the story of this remarkable fish, its 350 million-year sojourn in the waters of this planet, the capture of the first (East London) live specimen and the difficulties in reaching the second one lying on the deck of Hunt's schooner. Eventually the South African Prime Minister, Dr. D. F. Malan, authorized a military Dakota to take J.L.B. to the fish. It is a remarkable story, human, exciting, amusing and absorbing. It also highlights the efficiency of the South African Air Force, the co-operation of the Portuguese authorities, the graciousness of the French Governor of the Comores, the delight of the Comoran fisherman and above all the dedication of a scientist.

Each coelacanth caught means a considerable cash reward to the inhabitants of these once remote islands. Today, two decades later, every major and minor expedition in the Indian Ocean makes a special effort to visit the Comores and these islands are now attracting many tourists.

In 1954 with a hint of feminine guile I managed to include my fifteen-year-old son, William, as a member of our expedition to Kenya, Seychelles and the islands north of Madagascar. Six foot tall, well-trained in our work, and loving every minute of it, William made an ideal diving companion. For a month during this expedition we lived aboard a small forty-foot fishing vessel. We travelled close on 2 000 miles through some of the loneliest seas of the world, where the only land in the vast open Indian Ocean consists of a few tiny specks of islands north of Madagascar. Some are uninhabited and waterless, well off the beaten track. Those who have romantic ideas about coral islands should see these for they would soon be disillusioned. The only real beauty lay under the water.

For nearly six weeks we worked in the waters around Mahé, Silhouette, Praslin, La Digue, Mamelle, St. Pierre, the islands round Port Victoria and others of the Seychelles group. Then, with four hefty South African big game fishermen in the *Argo* – a trim, elegant yacht with flowing lines – the three Smiths set out in their fishing vessel to sample the fishes along the chains of islands south and south-west to Aldabra.

In some ways our voyage was a nightmare. A few hours out of Port Victoria, J.L.B. went across deck to tie some explosives more securely to the mast; huge seas from the south were buffeting the little boat. A sudden wave threw him backwards, he tripped over the coaming and was thrown down the hold, his back striking the edge of one of the steps. For some time his legs were paralysed; then came the pain. I spent as much time as the heaving boat permitted in trying to steady him in his bunk. He refused to return to Mahé and insisted that we continue the voyage. To complicate matters still further the engine stopped and for a long agonizing hour we drifted helplessly on the lonely ocean. Only when J.L.B. returned to South Africa, more than a month later, did X-rays reveal that one of his parapophyses had been broken. However, his insistence on continuing with the expedition and the resulting exercise had caused the bone to knit perfectly.

We continued our southerly voyage and visited many of the islands before reaching the Aldabra group. There we saw more fishes than we had ever imagined could possibly be gathered together in one area. Among other detractions we weathered very rough seas, a cyclone warning as well as a tremendous electric storm as we swung in complete circles round our anchor off the uninhabited Cosmoledo Island. We were almost consumed by hordes of mosquitoes as we tried to bathe in fresh water from the well of an uninhabited island, and on one occasion inside the Aldabra atoll, we succeeded in blowing ourselves up with a 50-lb. charge of gelignite. And we lived to tell the tale.

Moray eel. (Margaret Smith)

My son and I diving together worked as a team on this expedition. We had many experiences and adventures. At Shimoni we were examining a coral reef when suddenly we saw the head of an enormous moray eel protruding from a hole at the top of a huge coral head. It seemed to glare balefully at us. It had a neck thicker than my thigh, and it opened and shut its foot-long mouth, armed with three- to four-inch teeth, as it swayed from side to side. These eels are considered to be extremely dangerous to reef workers. Once they have hold of a victim they rarely let go. William urged me to poison it, but his father, warned by an innate sixth sense soon arrived on the scene and immediately ordered us into the boat.

Then there were the sharks. As we approached La Digue at the Seychelles, the sailors told us that a very large and bad-mannered hammerhead shark lived in the vicinity of the island. William and I decided to say nothing about this to J.L.B. as we badly wanted to work there. We agreed we would take special care and work in the channels between the coral reefs. Some time later I had forgotten the shark in my absorption in trying to net a very rare small fish, when suddenly something grabbed my leg. I automatically rounded on my attacker pulling my knife from its sheath, and prepared to plunge it into – my son! I do not know which of us got the greater fright, but he never again used that method to attract my attention.

Close to the island of Praslin in the Seychelles, where the famous coco-de-mer grows, we dropped anchor near St. Pierre, the most perfect little islet that I have ever encountered in Indian Ocean waters. Surrounded by large islands, this area was in a sheltered stretch of sea, and the large adjoining coral bank looked interesting.

In a couple of minutes we were in the dinghy over the reef, pulling on our flippers. Then quite suddenly a shoal of fish broke water near our boat. I saw brilliant chrome yellow and incredible blue and, with a shout, for I did not recognize them, I was over the side among them. I could hardly believe my eyes. I now recognized the shoal. We were in the middle of a school of the most bizarre of all the surgeon fishes, *Paracanthurus hepatus*. I knew the young well. They spend their time timidly hiding in coral, but here were the adults, fearlessly circling and soaring in a large shoal, their colours defying description. The yellow was of the brightest chrome, while the rest of the body was covered with an 'overwash' of an incredible mauve-blue which fades as the fish dies.

We had found a wonderful reef – sheltered water, prolific coral and gorgeous fish, unused to man and quite unafraid of us. In one hollow it looked as if a kind of fishy parliament was in session. All the fish assembled there, swaying gently with the water, as though they were discussing matters of great importance. In another spot some little rock cods stopped their play to look at us. Brown patches over their faces made them look like children disturbed at making mud pies, and they fell over backwards as they watched us float past.

Different fishes have different reactions, personalities and characteristics. One that never failed to amuse me was the thin, harmless, stick-like flutemouth, about six feet long. It would join in with shoals of other fish, try so hard to swim with them but be left behind each time. One large rock-cod at Assumption Island greeted us like old friends and greatly enjoyed William's patting and scratching him. We had to push him out of the channel where we were working, for he made such a nuisance of himself. Nine months later a National Geographic expedition had to cage this same rock cod because his curiosity interfered with their filming. Sometimes while swimming in open water, we would suddenly be surrounded by thousands of swiftly moving, brilliant, flashing, blue-yellow-pink *Caesios*. The water would be a whirling mass of them as thick as a cloud; in places we could hardly see. No matter how crowded they were, they never bumped into us. The movement was so rapid that we always felt quite dizzy and light-headed after they had passed.

There were many incidents in our underwater life – some amusing, some dangerous, some painful, some terrifying. The most interesting took place among the coral off lovely, crescent-shaped Assumption Island. The tide was rising fast, William and I had been hard at it for nearly three hours and were just picking up the last few specimens before returning to the boat when millions of tightly packed, small silvery whitebait suddenly came along and surrounded us. There is something in the fishy world that I have never been able to understand. Some fish seem to be born just to be eaten. The sardines that move up our Natal coast and cause such excitement are a case in point. They travel on and on, all the time providing food for frenzied fish, birds – and anglers! Whitebait are the same. Within a minute or two we were surrounded by huge fishes snapping up the silvery morsels as quickly as they could. William and I sat spellbound. We were merely parts of the scenery, for fish of 20 to 80 lbs. would come shooting straight towards us, and within inches of our masks, suddenly veer away. It was a never-to-be forgotten scene: kingfish, rock cods, barracudas, tunnies and even a shoal of the shy, exquisite, torpedo-shaped rainbow-runners, with blue bodies and golden stripes along their sides, rushed past. A goatfish nearby smacked its lips in an appreciative manner, while further on an eel had cornered some whitebait and was gobbling them up one by one. A fireworksfish used its long, elegant pectoral fins as a net to sweep the shoaling silversides into a corner and then slowly devoured them. We were guests of honour at a fish banquet.

The year 1956 was to be our last expedition to tropical East Africa, so we chose to return to Pinda Reef. Bringing all our experience and knowledge, we came back to the reef where we had started our tropical work. Our faithful Abdullah again headed our group of sailors. J.L.B. brought some professional fishermen who had come out from Portugal in a communally-owned line boat to fish for the Lourenco Marques market. He suspected that in the cold layer, 100 fathoms below the surface there should be fine eating fishes. He found them: magnificent snappers and large rock cods were brought to the surface, their bodies still cool from the temperature of their environment. One day this resource will be exploited.

I hired a slim, swift dhow or *lancha* that enabled me to work further down the reef. One day I noticed a heap of iron on the part of the reef that we were passing. In answer to my query, the sailors informed me that it was part of the wreck of the Greek ship at least half a kilometre to the north. I judged this to be almost impossible, so with my usual curiosity I ordered a change of course and my 'heap' proved to be fourteen ancient cannons encrusted with shells and barnacles. A small amount of rotenone brought out from between the long black spines of a tropical sea urchin living between the cannons eighty specimens of a small new fish, *Siphamia mossambica*, that we had described the previous year from only two specimens – one from Bazaruto and one from Kenya.

J.L.B. was very excited when I told him of my find, and the following day we stood together looking at the cannons. He reconstructed the tragedy: One calm, moonlight night well over 200 years ago a sailing ship heavily armed for protection against pirates or Arabs could probably see the land in the far distance, the captain little dreaming that a huge reef jutted out five miles into the ocean. Suddenly the keel ground into the highest part of the reef. Pandemonium certainly resulted, and to refloat the ship the desperate seamen jettisoned the cannons – obviously from one part of the deck. The ship obviously escaped being wrecked there, but I wonder if, without its guns, it managed to evade the pirates that frequented these parts.

And so, with our large collection of fishes, our photographs, colour sketches and notes, our experiences and our knowledge, we returned to our laboratory in Grahamstown. There we worked on the different fish families one by one. In 1963 we brought out the *Fishes of Seychelles*, really an illustrated catalogue of the most important fishes of the western Indian Ocean. In it we used black-and-white photographs for most of the fishes that were illustrated in colour in the *Sea Fishes of Southern Africa*. The coloured illustrations depict true tropical forms that never, or very seldom, reach our southern shores. Although our work merely scratched the surface of the vast fauna of the western Indian Ocean, its impact upon the international ichthyological community was considerable. Was it coincidence that within a decade this same area was investigated by numerous research vessels from various nations under the auspices of 'The International Indian Ocean Expedition'?

Chapter 8

Killer – the Zambezi Shark

No other shark in the Indian Ocean has claimed as many lives and caused as much havoc as the dreaded Carcharinus leucas – otherwise known as the Zambezi shark, Bull shark or Slipway Grey. Hardly in the 'heavyweight' class as sharks go – for they rarely exceed ten feet in length – the Zambezi shark is nevertheless a formidable adversary in its own environment.

The pack of sharks – there were five of them, varying in size from about 100 lbs. to 150 lbs. – circled the group of five divers in the water thirty miles north of Ponta Ouro in Portuguese, Moçambique. It was late afternoon and the visibility in this section of the warm Moçambique Channel was good. The temperature range, one of them recalled later, was 76–78 °Fahrenheit.

The five divers led by Natal veteran Tim Condon had been shooting fish all afternoon but had had little luck until a number of game fish swam into range. Conscious of the presence of the predators the spearos were at first a little cautious to start hunting, although the presence of shark in these waters was not unusual. What

A large maneater with remora-fish attached to its body in Indian Ocean waters. (Arthur Clarke: Seaphot)

worried this group of spearfishermen was that these were Zambezi sharks.

For a while the divers played tag with the fish, but when a 30-lb. 'couda swam into the sights of Noel Holliday's gun he let go, striking the fish clean through the head. His previous best on that holiday was a 67-lb. rock cod shot through the eyes only the day before. Noel lost no time in hauling in his line.

The sharks reacted immediately. From lazy, casual movements on the periphery of the divers' vision the pack suddenly became more active, closing the circle towards the divers and swimming with erratic twists and turns, the heads of some of them weaving from side to side. Here was trouble and Condon and his colleagues knew it. As the sharks pressed closer four loaded spearguns pointed outwards from Noel Holliday in the middle.

When the 'couda was about eight feet from its captor the largest of sharks made a determined dash towards the shot fish. With a thrust and a viscious bite it removed the entire tail. This was the signal for four remaining sharks to join in the assault.

In the greeny-brown murk of spilling blood and guts the attack turned into a frenzy with sharks making repeated sorties on to what was left of the fish. Moments later it became clear that it was not only the 'couda that the sharks were interested in as they lunged in pairs towards the spearfishermen. Two of them made a simultaneous attack on Holliday. With the butt of his unloaded gun he managed to repel one of the brutes while Condon literally hand-speared the other from lunging at Holliday's back.

At this point Holliday released his line and allowed the remains of the 'couda to fall to the bottom; for a moment the sharks turned their attention to a more likely target. Not to be outdone, Peter Yeld, another of the divers present, dived down towards another of the larger sharks busy scragging the last morsel. At a range of three feet he aimed his gun directly at the eye of the Zambezi shark and pulled the trigger. Fortunately – or otherwise – a tangle of steel wire caused the spear to slide harmlessly over the shark's back. At this moment the shark swallowed the last of the fish, barbed spear and all. The only problem at this stage was that the spear was still attached to Noel Holliday's gun. Only after being towed some distance by the shark, spear-shaft flapping at its side, did Holliday reluctantly concede that it would be the best for all to cut his line.

A few interesting points arise out of this attack, the first recorded incident of its kind along the Portuguese East African coast; there have been many recorded instances of single-shark attacks on bathers and divers before, but not in concert with others. On the same reef earlier in the day a number of fish had been shot. Although there were sharks in attendance there was no undue excitement on the part of these killers. No attempt was made to molest any of the divers even though another of the party, Noel Galli, at one stage swam with two large kingfish dangling below him laced to his line.

It is interesting that at the time of the attack there were a few species of other shark in the area, including Blackfin shark. Although these were excited by the antics of their grey cousins, they did not join in the fray. It was also noted that although one of the group of divers wore no body covering (the others were dressed in wet-suits), the assailants showed no particular predilection for him above the others; all five divers were subjected to vigorous on-slaughts by the pack.

As a maneater the Zambezi, as 'Zambe' as it is known to spear-fishermen, is a fearful proposition wherever encountered. Although restricted in the Indian Ocean mainly to the coast of Natal in Southern Africa and the Moçambique shoreline, the *Carcharinus leucas* – otherwise known as the Shovelnose Grey, Slipway Grey

The Zambezi shark (Carcharinus Clucas) is an aggressive killer and as unpredictable as the weather. (Oceanographic Research Institute, Durban)

(because they are found near the slipways of whale processing plants) Bull shark (Australia) Cub shark and Ground shark (U.S.A.) and Lake Nicaragua shark – is regarded as the maneater responsible for more than 95 per cent of all shark attacks in Southern Africa. A few of these onslaughts have taken place in fresh water, in some places more than 300 miles up the Zambezi river in Central Africa. Of the twenty-two shark attacks recorded by Dr. David Davies in his book *About Sharks and Shark Attack* (Shuter and Shooter; Pietermaritzburg 1963), all but one are listed as having been inflicted by this dangerous, though not overly large predator; they average in size between six and ten feet although many smaller 'Zambes' are known to have harassed divers, particularly off the north Natal coast.

The pattern of attack is rarely the same. Active spearfishermen recall that attacks have taken place for no apparent motive in waters that have ranged from crystal clear to murky in depths ranging from a few feet to deep water reefs. These are usually warded off with spearguns. Most attacks are limited to inshore areas, invariably around polluted river estuaries or harbours. What is known is that attacks are far more likely to take place in the late afternoon and that a water temperature of 75 °F or more is critical in this connection. There is also more likelihood of attack if the water is dirty.

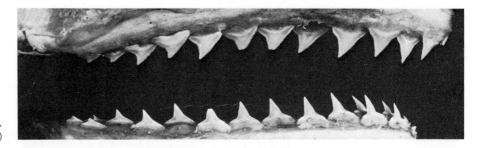

Teeth of Zambezi shark.
(J. L. B. Smith)

Clearly the variety *Carcharinus leucas* is one of the more aggressive and foreboding sharks to be found in any ocean. Unlike its sleeker and more graceful cousins, the Black-tipped shark (*Carcharinus maculipinnis*) Blue Pointer (*Carcharodon carcharias*) or Tiger shark (*Galeocerdo cuvieri*), the 'Zambe' is an ugly beast with a wide head and extremely short, proudly-rounded snout. Its appearance, in the words of the late Dr. Davis is 'heavy and its form compact'. It has a prominent triangular first dorsal fin and small eyes. The upper teeth are coarsely serrated and broadly triangular; lower teeth serrated and triangular but more slender than the teeth of the upper jaw.

Superficially the 'Zambe' resembles its distant relative the Dusky or Lazy Grey shark (*Carcharinus obscurus*) which is also common in the waters of the Moçambique Channel, but is readily distinguishable by the absence of a dorsal ridge.

Nor is this species adverse to attacking other sharks should the opportunity present itself. One large specimen which was kept in captivity for some time by the Durban Oceanographic Institute was eventually destroyed after it had attacked and eaten a number of other valuable sharks with which it was sharing a tank. The decision to kill this shark was taken after it had taken an unusually active interest in some of the divers accustomed to entering the tank for feeding purposes. 'Willy', the shark in question, had become a major attraction at the oceanarium and scientists associated with the Institute concluded that killing it was the only measure left to them to ensure safety.

There are also instances on record where Zambezi sharks have attacked boats and canoes. One such incident occurred in January, 1963 when three canoeists on the Limpopo, 120 miles from the sea near Guija, in Moçambique, repeatedly had their canoes bumped by a large Zambezi shark. Over eight feet in length the shark made a number of charges until it eventually lashed through the fabric of one of the boats. Its occupant Jopie Overes a student at the University of Cape Town was obliged to paddle as fast as possible to reach the shore before his canoe sank. He only just made it with the shark in hot pursuit.

Reef shark on patrol.
(Seaphot)

Little is known about the life span of sharks except that some specimens have been kept in captivity for more than thirty years. One particularly husky specimen of 'Zambe' was kept in the Durban Oceanarium for six years. It is known that these sharks take some time to reach sexual maturity. Young Zambezi sharks have been kept for varying periods of between three and five years without reaching sexual maturity. Like the Hammerhead (*Sphyrna lewini*) and the Dusky shark, the Zambezi shark has to keep continually on the move day and night. All those in the Durban Oceanarium have been observed swimming throughout their period of captivity. The only time this species rests on the bottom is when it is affected by illness or injury and unless a diver is immediately dispatched to 'walk' the shark about, it dies within a short time due to lack of oxygen. The continual movement of the shark allows oxygenated water to circulate through its gills.

One of few known injuries incurred by divers in African waters (there have been no divers killed by sharks) resulted from an attack off the Zululand coast by a Zambezi shark. Another incident occurred soon after in the early sixties when Len Jones, a well-known South African undersea enthusiast was lightly injured by a Blue Pointer shark while diving north of Durban. The Blue Pointer had gone for the fish on Jones' belt.

In the Zambezi incident, the first of its kind in southern waters, a local diver, Clive Passmore entered the sea at Umdloti on his own. He had just arrived on the beach when he saw three divers emerging from the water after spearfishing off Selection Reef a point about 150 yards from the shore. Visibility, Passmore was told, was about twenty feet and the sea temperature was just over 70 °F.

Passmore recalls that he had not intended diving alone but having come that far and conditions being fair, he did not think he would be in the water for long. He was not.

I ducked through the shore dumpers and swam straight out through the foam barrier. Loading up at once I headed for the north corner of Selection Reef and began hunting. When diving alone I have always been in the habit of keeping a sharp lookout below, behind and to the sides of me to make sure I'm not being followed.

I had been at the reef about five or ten minutes and had just made a round-the-clock check when, quite suddenly and without warning I received a terrific jolt on my right elbow which shook my whole body. Jerking my arm back sharply I turned, just in the time to see the mauvish-

Lower tooth of Zambezi shark. (Margaret Smith)

grey back of a six-foot Zambezi shark turn and streak away into the murk.

A full second went by before I realized that I had been bitten and then I began to scream into the water after the shark – almost hysterically, I think. The shark was gone by then leaving no trace save a thin wisp of blood trailing from my arm. I remember not daring to look at the wound for fear that I might panic and make matters worse. Holding my gun at the ready in case it should return, I turned and swam flat-out for the shore, the whole time twisting left and right and back.

I felt no pain at this stage – just a sense of disbelief that this could really happen to *me*. Considering the size of the wound there seemed to be very little blood in the water.

I reached the foam without seeing the shark again, then plunged into the zone of bubbles and zero visibility with my heart in my mouth. Feeling my feet strike ground was a terrific relief and with complete disregard for the pounding I might get on the sharp rocks from the next swell, I clambered up over them to the beach.

The people on the beach had seen by then that something was amiss and came running from all directions. I turned to my fiance and said simply: 'A shark got me' and sat down on the sand. By this time the wound had become fairly painful and my right hand was a little numb. I was made to lie down for a while and was then taken to hospital where, quite suddenly, my lone Sunday afternoon dive seemed to have become of interest to a great many people.

Looking back on the incident Passmore could think of little, if anything, that might have provoked the attack. True enough, he recalls, a fish had been shot but that was a while before. The temperature was definitely below average and he was not wearing a bright bathing costume. Nor was he carrying fish on his belt. In short, the attack was 'against the rules'.

The fact that I was alone in the water had little, I think, to do with the fact that the shark (it was later confirmed by the bite as a Zambezi shark) attacked. I think it would have attacked even if there had been another diver with me. It would have mattered a great deal, however, if the Shovel Nose had returned to finish the job it had started. In that case I don't think I would be here to tell the story.

Chapter 9

Agulhas's Tanker Graveyard

With the closure of the Suez Canal in 1967, a formidable array of shipping passes around the Cape each day. Because of this flow and occasional adverse weather conditions, particularly during the winter months, a number of these ships – including several giant oil carriers in excess of 100 000 tons have come to grief. Diving on these hulking wrecks adds a new dimension to undersea ventures at the southern tip of Africa. **Brian Rees** *recalls.*

The reef before us – part of the Agulhas Bank at the southern tip of Africa – started to take shape and form as we swam nearer; man-made form. Our startled eyes blinked as a massive propeller, almost as big as a house, glinted at us through the diffused late-afternoon sunlight. It stood perfectly upright and was connected only to a ragged section of the ship's hull of darker hue and of about equal height lying on the floor of the ocean. We swam on rapidly, our excitement registering in the abnormal flow of air passing through our scuba demand-valves. For a moment or two we were peremptorily halted by strong eddies around the huge black artificial reef caused by twenty-foot swells which rolled in a steady pattern across the surface of the sea sixty feet above our position.

The undersea terrain in the Southern Indian Ocean is different from the tropical areas. Here the sea is colder and kelp replaces coral. (Al Venter)

Seals often accompany
divers in southern waters,
especially around Cape
Agulhas and further west.
(Alex Papayanni)

Mike Clark and I conversed rapidly in sign language and took turns to pose against the angled piece of yellow metal so that each of us could record the discovery on film. Without this proof of our unusual find we were sure no one would believe us for it was April 1, 1972 and this was the sort of trick any diver would love to play on his knowledgeable wreck-hunting friends ashore. How else could we make them believe what we had found out there; a five-bladed copper-bronze propeller about eighteen feet in diameter.

But there were no April Fools where we were diving, only sharks, and this wreck was for real. Behind the propeller lay the rudder, trim and elongated and almost as big as the prop. It too was barely damaged, lying on its side against the reef; like the propeller it showed only a few minor depressions along the edges.

Still underwater and taking stock of what lay before us, my mind attempted to reconstruct the disaster which had caused all this. We were five miles from the shore and just off one of the busiest sea lanes in the world since Suez was closed five years before. Shortly before we had dropped our anchor we had passed a freighter of some 20 000 tons to seawards not more than a few hundred yards off our port beam.

The stern of our wreck pointed due west; the fractures on what remained of the hull indicated that this piece had been torn from the rest at speed. The balance must surely lie to the east, perhaps in deeper water off the deep water reef we had come to explore.

Mike's thoughts and mine were obviously running along a similar tack. Together we moved away ahead of the propeller and down the shaft which ended close to our own anchor. We followed its direction, and about thirty feet further along found it again, and with it, a warehouse-sized mass of tubes, plates and valves.

The sight of a six-foot shark lying dead between a set of twisted tubes brought us rapidly back to reality. This was a no-man's land for the likes of an amateur diver; terrain dominated alternately by game fish – marlin, tuna and tunny and sharks. We were also aware that five miles off shore in the warm Moçambique current the sharks we would encounter were oceanic sharks, the sort that rarely allow for a formal introduction. At that moment, a large white shape loomed out of the blue about forty feet away and swam leisurely towards me. Armed only with a camera, I felt ominously exposed and looked round for Mike who was armed with an aluminium shark prodder which he had taken down in anticipation of just such an event. By the time I had located him, a 'ragged tooth' shark of about ten feet was alongside us. A moment later it had disappeared as rapidly as it had appeared, away into the gloom; too soon, in fact, to leave an impression on our films.

I had had encounters with ragged tooth sharks on previous occasions on this reef, usually when spearfishing for yellow-tail amberjack, but they had never troubled me with more than an occasional, disinterested 'buzz'. Large, ugly and slow moving, the 'ragged tooth' has been known to take fish, but never a diver.

We swam on, over what appeared to be the engine-room, all parts of which were covered in a centimetre or two of weed and marine growth. It ended suddenly in a patch of sand. We dropped to the bottom again in an attempt to identify some part of it.

The author explores a wreck in the Agulhas area. (Mike Barhouch)

Having been submerged for almost an hour, I switched over to scuba reserve, and indicated to Mike we must surface soon. Looking around for some memento, I picked up a length of brass tube, and swam back along the side of the engine-room. In the lee of the huge mass of crumpled machinery I noticed a small brass cage lamp protruding from the mass and pulled at it, never expecting it to be loose. The item came away easily and it was my surprise to find I had extricated it in good condition which made for an excellent

The undersea terrain in False Bay is at times stark and bizarre; it has little of the beauty of a tropical area, but nevertheless provides many attractions. (Al Venter)

memento of our visit. Even more unexpected was the brass manufacturer's plate on the lamp – 'Morio Denki Co. – 1955'.

During the latter part of the dive my mind had again been turning over possibilities. The reef we were on was well known to both professional and amateur line-fishermen who worked it regularly throughout the year for yellowtail. Only in recent years had spear-fishermen ventured out, and we were perhaps one of the first group of scuba divers to have dived there.

Up to the moment of finding the lamp, I was sure that what we had found was part of a war-time wreck, blown apart, perhaps, by one of the German U-boats which were active in the area. But the lamp showed clearly the wreck was post-1955, and in such recent times I had no recollection of so large a ship foundering in this area.

Back in the boat again we found our colleague Ernest Brinkmann suffering severely from sea-sickness brought on by the heavy swell. Our other 'buddy' Terry Attridge had his speargun out and had accounted for two fair-sized yellowtail from one of the many shoals which passed over the reef while we were down. On hearing our story Attridge promptly donned Ernest's scuba set whereupon he and Mike returned to examine the wreckage again.

While they were down I attempted to take precise bearings of our position, but this was no easy task with a predominantly flat terrain ashore and few features other than the Agulhas lighthouses showing themselves. South-east of our position was the reef's pinnacle, indicated by the flotilla of small boats fighting for position. It was for this reason we had moved away, not wishing to become entangled with anglers' lines.

The sea was picking up rapidly so as soon as Mike and Terry had come onboard again we headed for shore. Three times again we tried to reach the wreck that weekend, but each time we were forced to turn back by gale-force winds. The site of the wreck was no place for a twelve-foot inflatable boat.

The first three days of the new week passed quietly. Finally Mike's curiosity got the better of him, and he visited the shipping editor of a Cape Town newspaper, the *Cape Times*, to find out whether he had a record of recent wrecks along the Agulhas Bank. We had decided among ourselves to say nothing to local divers until we had identified the wreck. George Young, the shipping editor, wanted to know more, and asked Mike to send me along with the lamp so that he could examine it. He promised to put a small column in his shipping page the next day in case someone could aid in its identification.

This editor, however, had other ideas. The following day we found ourselves front-page news with a four-column spread of our find in the *Cape Times* and banner headlines on posters all over town – 'DIVERS FIND MYSTERY WRECK!' So much for our security measures. On realizing that our discovery had been made on April 1, the same editor had second thoughts the following day. He was almost convinced that he had been hoaxed by a bunch of divers. Then, much to his relief, our first pictures came off the spools.

The next week gave us a foretaste of what celebrities must go through every day of their lives. Telephone calls, reporters, requests from Lloyds of London for more details, more articles in the paper, and hosts of plausible and absurd new theories as to what the wreck could have been, and how it got there. We were taken aback by the reaction which we thought was completely out of proportion to the chance find we had made. The most far-fetched theory of all was that it was the hull of the tanker *Silver Castle* which had gone down off the east coast 600 miles away. It was maintained that the submerged hulk had drifted down the coast underwater, to lodge itself on the reef. Or so the 'experts' said.

With this blaze of publicity, we thought we owed it to the public to go back and make a further search of the area for more clues. But it was not to be, for the winter storms set in early. Two attempts were abandoned after a hectic 140-mile drive down from Cape Town on successive Fridays after work.

It was some weeks later that Lloyds' agents supplied the answer, having traced the manufacturers of the lamp. They had first suggested we go back to the wreck and find the number stamped on the boss of the propeller shaft. When we explained that it was well nigh impossible to find a one-centimetre number on a one-metre diameter boss covered in marine growth with a heavy ground swell, they discarded the idea. This view was telexed to their intelligence branch in London.

When Lloyds announced that what we had found was most likely the remains of the *Wafra*, a large tanker which had gone aground off Agulhas some months earlier, public interest had waned and there was little reaction. The *Wafra* was finally freed from the rocks and towed some hundreds of miles out to sea where it was sunk by South African Air Force *Buccanneers*.

Local divers who knew the reef better than ourselves had suggested the *Wafra*, but having discussed the matter with Jim Hartly, Cape

The *Wafra* seen from the flight deck of an Air Force helicopter as she sits fast on the rocks off Cape Agulhas. The tanker was later pulled clear of the rocks.
(*The Argus*)

Town doyen of the salvage world who had been on the tug which towed the ill-fated vessel off the reef, we discounted the theory. He was of the opinion she had been towed off intact, though this was never substantiated prior to her eventual sinking.

But the *Wafra* had been big news at the time. The first distress call was received early February 27, 1971, when the Captain of this Liberian-registered ship of some 50000 tons, owned by Getty Tankers, reported a flooded engine-room. His radio message reported that the ship was drifting without power about eight miles off Cape Agulhas. She had been travelling with a full load of crude oil from the Persian Gulf for delivery to Cape Town.

With memories of two previous near-tragedies still fresh in the public's minds when giant tankers had stranded within sight of Cape Town and threatened to spill tens of thousands of tons of oil into the sea, the words 'another Torrey Canyon' were on everyone's lips. Luckily disaster was averted in both these cases – the stranding of the French *Sivella* at Green Point on February 3, 1968, and the *Kazimah* on Robben Island on November 24, 1970; but both ships were freed just before they were broached by heavy seas.

The *Wafra*, however, was drifting in an area of dangerous shoals where the continental shelf extends 400 miles south of the shoreline, and where mountainous seas have claimed many a good ship before her. All but eleven of her crew were taken off by the Russian tanker *Gdynia*, and two deep-sea salvage tugs from Cape Town set out immediately to render aid. The towline between the *Gdynia*, and the *Wafra* soon parted in rough seas and by the time the tugs reached her she was fast on the Five Mile Bank.

Two days later her tanks burst. A river of oil streamed from a huge gash which could easily be seen from the air, to produce a slick more than sixty miles long. By this time the American salvage specialist Harry Millard had taken charge of salvage operations and two large compressors were landed on the decks by Air Force helicopters; air was pumped into the ballast tanks to lighten the ship. Meanwhile two German salvage tugs, the *Baltic* (5000 HP) and the *Oceanic* (17500 HP) were battling to move the *Wafra* from the reef. The following day the number of leaking tanks had increased to six and two small fisheries boats tried vainly to cope with the massive oil slick which had reached the shore in many places.

On March 3, less than a week after she had grounded, the *Wafra*

The *Wafra* burning furiously
south of Agulhas after being
bombed by South African
Air Force *Bucanneers.*
(*The Argus*)

was written off as 'unsalvable' by Mr. W. F. Averill, president of
Getty Tankers, who had flown out to supervise the operation. It
was then decided at Government level that the ship should be
destroyed by the Air Force. But a dispute then arose between the
captain of the salvage tugs and the owners, so while this went on
more salvage schemes were devised. It was only five days later,
with the help of two more Cape Town tugs, that the *Wafra* was
finally freed on the spring tide. Reporters noted that her stern was
'riding low' in the waters, but the crippled tanker was towed south
for three days, and finally sunk in 2 000 feet of water after the South
African Air Force had managed to set the oil alight with incendiary
bombs.

There, to the relief of many South Africans, the *Wafra* died. Until
we stumbled on the propeller no one was aware that while being
pounded on the reef, her back was breaking; both engine-room and
propeller remained behind when the rest of the hull was towed
away. Evidence of the lack of damage to the propeller and the large
amount of fractured rock in the vicinity, supports this theory, as
does the lack of any further sign of wreckage. The propeller is still
there, valued at about £5 000 and guarded by some of the most
treacherous seas and currents in the world.

This disaster, however recent, was not the last. Tragedy, this time
with heavy loss of life, struck again on August 21, 1972. Only twenty

miles east of the reef where the *Wafra* stranded, a mammoth Liberian tanker of almost 100 000 tons, the *Oswego Guardian*, collided in thick winter fog with the Greek-owned *Texanita*. The *Texanita*, a ship about the same size as the *Oswego Guardian*, exploded instantaneously. Of the thirty-five men on board twenty-six were lost although six merchant ships together with boats of the South African Sea Rescue Institute besides aircraft searched for survivors for two days. The *Texanita* was returning to the Persian Gulf in ballast with her tanks uncleaned and probably full of inflammable gas, while the *Oswego Guardian* had a full cargo of crude oil onboard destined for Europe. Only the bow of *Oswego Guardian* was damaged, so that after survey and fairly extensive repairs in Cape Town she was allowed to continue to her destination.

The horror of this tragedy was recorded in full on the radar screens of a number of other ships in the vicinity, but unfortunately for those involved neither ship was aware of impending disaster until the very last moment. An officer onboard the Norwegian vessel *Thorswave*, about six miles from the scene of the accident, witnessed the full preliminaries to the collision, and was powerless to stop the two green dots upon his screen merging into one. The explosion was heard sixty miles inland, and the mushroom-shaped cloud which enveloped the scene immediately afterwards caused some local residents to take cover for they thought that World War III had started.

A sequel to this tragedy was recorded two months later when in October 1972 the South African Navy Hydrographic vessel *Protea* fixed the position of the hulk off Cape Infanta. The *Texanita*, it was found, lay in 200 feet of water, and she had come to rest on the bottom in an upright position. Traces of oil were found in the sea nearby, but there would certainly have been far greater pollution had she been sunk with a full cargo.

The *Texanita* lies deep, but not too deep. Because her position has been chartered there are enthusiasts 'down south' who are preparing to dive on her. They are only waiting for the weather to clear. Meanwhile one of them has already built a small metal shark cage in which to go down for he, like the others, accepts that it will not be an easy venture.

Chapter 10

Sharm-El-Sheik – Red Sea Paradise

The Red Sea has always intrigued divers. Not only is this remote tropical area considered among the best of diving haunts in the world, it has often proved the most inaccessible. The Egyptian Government imposed bans on diving off their coast after the 1967 débâcle; and it was five years later that the opposite coast was opened up by the Israelis. **David L. Livingston**, *an American, writes of his experiences in the area.*

Clad in my warm-water wet-suit, full scuba gear, and an underwater camera in hand I began crossing over the coral reef to the diving spot. Behind me lay the desert. As I looked overhead, the cloudless blue sky came momentarily to life as several Israeli Air Force F4 Phantom jets shrieked by. The place was Sharm-el-Sheik, located at the southern tip of the Sinai Peninsula on the Red Sea.

Because the Sinai is still classified in Israeli terms as 'occupied territory' – one of the results of the Six-Day War – I was obliged to travel through Eilat to reach the place. But the drive is an interesting one; the Israeli-built highway from Eilat to Sharm is good and the trip takes about four hours under normal conditions although no journeys are allowed at night. The road meanders between the outer edges of the Sinai and the ever-widening Gulf of Aqaba; a stimulating experience and a prelude, perhaps, to the breathtaking scenario which unfolds the moment one enters the water at this natural undersea paradise known to increasing number of divers as Sharm-el-Sheik.

As a diver and underwater photographer, it was my intention to explore the coastal areas between Eilat and Sharm in the Gulf and also to dive and photograph the Red Sea adjacent to the southern tip of the Sinai. In these areas underwater conditions are superb. At places such as Ras Nasrani, Ras Muhamed (all Sharm locations) and Dahab, there are coral reefs equal to the best anywhere else in the world; coral life of all types and fish of a staggering variety. Other spots along the coast provide more than enough excitement and beauty to numb the emotions of even the most experienced diver. My last visit to the Red Sea was in December of 1972 and I

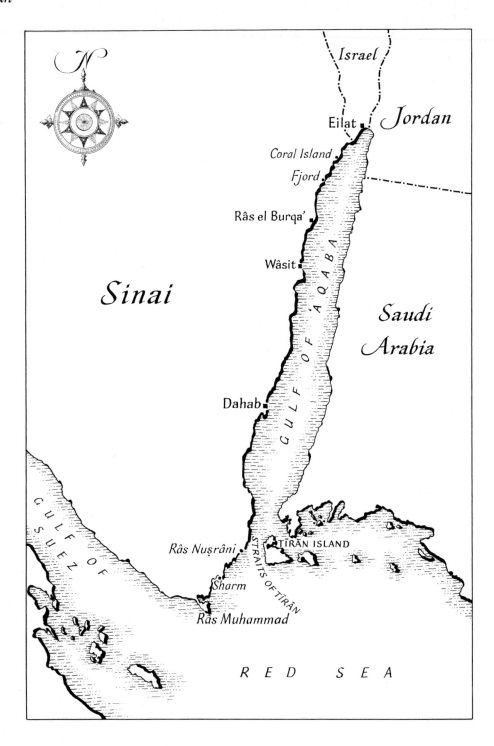

was joined on the expedition by Dr. Donald A. Frey of the Oxford Research Laboratory for Archaeology, Oxford, England. Together, Don and I explored this area extensively.

Our adventure began in the Gulf of Aqaba opposite Coral Beach,

Garden eels dancing on the sea floor off Coral Beach. (David Livingston)

just South of Eilat. Directly in front of Aqua Sport Diving Center, roughly thirty metres from the beach there is a colony of several hundred 'garden' eels. These eels are not common and offer something very special for any diver to see; most are burrowed in shallow water from twenty to forty feet, gathered together in communes at the top of a ridge with a gently sloping sandy bottom down to 100 feet.

Because the eels frighten easily it is not possible to approach too close to them without using a blind. However, the water is usually clear and it is easy to observe their swaying, dancing motions from a distance. When a diver comes too close to one of these eel communities they slowly retreat into their lairs on the sea floor.

Silver grey in colour and less than an inch thick the eels appear to sport a type of mane on their neck near the head.

But the best diving in the area is to be found a few hundred yards further south at a giant coral outcrop called Moses Rock. The area is renowned for some of the best diving in the Gulf. There are numerous varieties of undersea flora and fauna and the fish here are quite docile, almost trained, as the majority are attracted rather than frightened by divers. Frequently I glanced down towards my fins and noticed several butterfly fish swimming around my legs. It was as if I were part of the coral structure. The green and black parrot fish seemed to follow me as I swam around the rock and for a while

Butterfly fish on a giant coral outcrop near Moses Rock, Eilat. (David Livingston)

my attention was detracted from a wealth of angel fish, wrasses, clown fish and corals in my immediate environment.

Some of the best dives at Moses Rock are made in the early evening. At this time of the day the delightful feather stars become visible as do the cocky turkey fish, both in increasing abundance. Those at Eilat were the largest turkey fish I have ever seen and mostly in a hue of brown or dark maroon; Moses Rock seems covered with them and it is an experience to watch dozens of these creatures, gracious yet dangerous, move slowly about the area.

Moses Rock is also an exceptional place to dive for visual photographic impact. The corals in the area reflect mostly yellow hues but there are also dozens of shades of blue, red and turquoise.

Moving down the coast of the Gulf a few miles, the next diving spot is Coral Island a small off-shore land mass, three hundred yards

Lettuce coral on the sea floor at Coral Island. (David Livingston)

from the beach. On the island stand the remains of a fortress built by Saladin in the 12th Century. The barren desert-brown island in the middle of a sky-blue sea with unlimited visibility below the surface has created excellent diving possibilities for enthusiasts.

The best diving is on the far side of the island and at the southern tip. A small boat is available to take you there, but the swim is short and enjoyable. After reaching the island, it is necessary to cross over it via a path to reach the other side.

As soon as one crosses narrow reef and submerges, large and strange coral formations loom on all sides. These coral heads are scattered along the sea floor from shallow depths to over a hundred feet and the diver with imagination can have a field day; I remember distinctly that one formation reminded me of a camel being ridden by one of the local Bedouins.

There has been some work done of an archaeological nature on and around the island because of its historical importance. Relics of pottery have been discovered in the water and from reading through a report made on the subject, I suspect that there is still much waiting to be found. For those interested and wishing to extend their activities it is possible to do some underwater archaeological exploring here. Finding bits and pieces of pottery from the time of Saladin is always fascinating. The coast of the Gulf of Aqaba down to Sharm boasts many such picturesque and exciting diving locations.

With the Sinai mountains behind one and a clear view of the Red Sea – the Gulf of Aqaba and Saudi Arabia lie across the bay – the diver has the opportunity to enjoy coral reefs and sea life under a desert environment which is very different from what one finds elsewhere. Such spots as the Fjord, Ras-el-Burqa, Wasit and Ras Abu Galum all have their attractions. These areas are also known for their varieties of moray eels but it was while diving at the Fjord and at Wasit that two adventures unfolded.

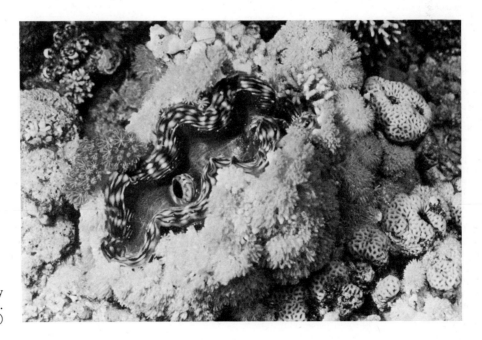

A fine triadacta clam fully opened off Ras-el-Burqa. (David Livingston)

The Fjord, nine miles south of Coral Beach, was so named because it does in fact resemble a Norwegian Fjord. Its natural beauty is vested in the turquoise-blue water and a visible reef surrounded on all sides by the barren brown and rugged hills of the Sinai. From the approach to the Fjord, the road has to cross the top of a pass; the first view from this height is breathtaking.

I was diving here with an Israeli friend and after enjoying a very pleasant and uneventful dive looking at coral and fish, we were on our way in when I saw swimming very slowly along the sandy bottom what I believed to be a sea snake. It had white and black stripes and was about three feet long. It reminded me very much of the sea snakes I had seen in the waters of the Fiji Islands in the South Pacific and I knew that this was a very poisonous creature. I succeeded in drawing the attention of my buddy. He must have thought the snake was an eel because he proceeded to swim rapidly towards it with outstretched hands.

At first I thought he was having a bit of fun, but as the drama continued, I saw my friend actually pursuing the snake which was now moving strongly in the opposite direction. As I continued watching with an increasing sense of disbelief and amazement, my friend reached out his hand, grabbed the tail of the snake and let go of it just as it was about to strike. The snake darted into a nearby coral hideout and that was the last we saw of it. Only afterwards did my friend admit that he thought it was an eel.

Back in Eilat, we recounted our story and it was the general consensus that sea snakes do not live in this area. Everyone who was prepared to voice an opinion said it was an eel, but I am still certain that it was a snake. Regardless, the event provided sincere moments of anxiety and tension, besides many laughs afterwards.

While diving at Wasit, about an hour's drive southwards from Eilat, the same companion and I were enjoying a rather remarkable photographic dive. Our air was getting low when suddenly, in eighty feet of water, we found stretching away below us a fantastic coral reef drop off, the best that we had seen so far. The water was perfectly clear with at least 150 feet of visibility. We both checked our decompression and air gauges and decided we had air and time for a quick plunge over the edge. When we finally reached the bottom our depth gauges read 120 feet.

But it was the marine life which attracted us. There was a never-ending abundance of coral and large fish. I immediately busied myself shooting pictures of the area because I knew that time was limited. With one picture remaining for something unusual on the way home, I signalled my buddy and we began ascending back over the reef. When we reached the top of the reef again at eighty feet, I looked straight up and saw some five feet away, a single barracuda about five feet long. Quickly, I snapped the last remaining picture on the spool and motioned to my friend. The barracuda, which had been swimming past, suddenly changed course and swam back towards us very slowly, turning only for an instant to display a mouth with two long rows of razor sharp teeth. It followed us all the way in to shallow water, keeping station never more than a few feet away. For a while, I recalled afterwards, I thought the predator would present problems; but it went its own way in the end.

I have referred to three other places as my favourites for diving. Of these three, Ras Nasrani and Ras Muhamed are situated in the Sharm-el-Sheik area, while Dahab is roughly an hour's drive to the north. Diving is excellent at all three points.

It is impossible for me to forget my first dive at Ras Nasrani. After diving off the coast of the Gulf of Aqaba for two weeks in the summer of 1972, I eventually worked my way to this area located directly across from Tiran Island, where coral reefs border on the now-famous Straits of Tiran. Looking in the direction of Tiran Island, one side forms the mouth of the Gulf opening to the Red Sea; on the other side, also clearly visible, is the Gulf of Aqaba. Far off in the distance lies the Saudi Arabia shoreline, hanging suspended in the haze almost like a mirage caused by the hot desert sun.

The Straits of Tiran are deep, extending to over 1 000 feet in many places, and a few spots to more than half that again. It is the same at the coral reef drop-off at Ras Nasrani. Suddenly one drops from a depth of about seventy feet at the base of the reef to a 1 700-foot abyss in the middle of the very narrow navigational channel. When the diver gets a first look beneath the surface here, it is an entirely different world from any of the other nearby diving spots. The water is warm and clear; there is an abundance of large fish, sharks and coral.

During my dives at Ras Nasrani, I always sighted sharks, usually white-tips about six feet long. There is so much in the way of food along the reef for these creatures that they appeared to be well fed and rarely bothered divers. In fact, on several occasions I approached sharks to take their pictures and they left before I could get within range. I had not experienced this type of shark behaviour before but it is verified in part by local stories because there is so much marine life about the sharks are wary of anything large, including divers.

Regularly, while diving at Ras Nasrani, I would position myself someplace along the reef, usually about eighty feet deep and look upward. Due to the unusual clarity of the water, it was possible to follow the movements of all fish between my position and the surface; from large tuna to the smallest golden blue-eyed reef fish which breed prolifically in these tropical waters. There is usually a current in the area because it is near the entrance of the Gulf to the Red Sea; so it is best to start out any dive swimming against the pull.

Among the many fish that inhabit this reef, the clownfish (*amphiprions*) are most astonishing. In this area these fish grow large and are often aggressive. It is not unusual to see a clown fish leaving its anemone to attack a would-be intruder. This is also the home of some fairly large spotted coral trout or grouper which make excellent subjects for photography.

Not far from Ras Nasrani on the south side of Sharm-el-Sheik is a point jutting out into the Red Sea called Ras Muhamed. From here I could see the coast of Egypt lying mysteriously behind the island of Chadoin and off to the right the opening of the Gulf of Suez.

At Ras Muhamed, recently declared a nature reserve by the Government of Israel, the most prolific sea life exists. It is here that I found the largest groupers, the most shoals of angel fish and by far the most fascinating assortment of other fish, corals and sponges, all of which provided ideal subjects for both day and night diving. To identify locations in the Ras Muhamed area the desert road is marked off, using barrels with numbers painted on the side. At the very end of the road, all the way to the point and past the first barrel, we found the best conditions for diving.

Access to this spot is good; a swim of about 100 feet through waist-deep water to the reef. Once on the other side of the reef the real submarine vista takes shape. Moving south-west from this point one enters an area where large amounts of black coral can be seen in depths as shallow as seventy-five feet.

After making several dives here, Don and I decided to enter at the

Grotto fish in the southern
part of the Sinai area.
(David Livingston)

same point, but swim farther on to the west in a bid to take shark
pictures. As we were snorkelling out to the predetermined dive
site, I spotted two of the elusive white-tips beneath me. I attracted
Don's attention and moments later we were heading downwards
ready to photograph come-what-may. But as soon as we approached
the two sharks took off.

We surfaced again and snorkelled back to our cross-over spot on the
reef. Once back in shallow water, we stood up and began talking. I
noticed that Willy was frantically waving his arms and yelling something at us, but neither Don nor I could make out what he was
saying. We continued talking for a short while longer and then
began walking and splashing back to the beach. It was then that
Willy started running into the water with a large tent pole in his
hands; this action really had the two of us confused. Willy drew our
attention to our left and there, twenty feet away was a five-foot
black-tipped shark swimming in lazy circles on the surface with its
dorsal fin out of the water. With all our splashing and irregular
movements the shark, which had evidently followed us over the
reef, was now looking for the way out.

But the most interesting part of this reef for diving is off to the left in
about thirty feet of water. In this area the diver encounters the first
of a series of underwater grottoes. Most of these undersea caves are
reminiscent of a subterranean journey, with many passages leading
through to different caverns, although the grottoes are not deep.

There are more grottoes in between the first and second barrels, about a mile from the point at Ras Muhamed; it was here that Don, Willy Halpert of Aqua Sport in Eilat, and myself camped for several days while diving. What impressed me most while exploring there was the sensation that I was diving into an Israeli underwater version of the Caves of Dios in Greece. There is also much black coral about in deeper waters.

Our final dive at Ras Muhamed was at night. This was Don's first night dive and after explaining how the lights worked so that I could take pictures, we entered the water to explore the underwater grottoes by artificial light. Since we assumed we were familiar with the area, we thought we knew what to look for; but with the sun well down, the grotto took on a different character altogether. I was amazed at the variety and profusion of yellow, red and pink soft corals which were not nearly as apparent during the day. At night, everything our lights touched seemed to come alive. Many of the growths supported a type of feather star.

To the uninitiated the grottoes at night are almost like something from the realm of a science fiction novel. There was enough ambient light from the moon and stars to silhouette grotto forms with their rather austere openings back into the reef. The scene was spectacular and eerie; at times it was quite frightening.

I found it impossible while in Israel to separate the aesthetic grace and proportion of the underwater world from the primeval beauty of the Sinai Desert. The area around Sharm-el-Sheik was, in retrospect, almost a fantasy reaching far under the sea.

To me, diving at Ras Nasrani, Ras Muhammed and other areas near Sharm such as Shark Bay and Dahab, offers the ultimate in an intriguing environment where many unique forms of life are present in their natural state; they are as yet untouched and untroubled by souvenir-seeking tourists. Government controls intend to keep it that way. Also, the excitement and the action are there to the full. It is apparent that my feelings are shared by increasing numbers of foreign and Israeli divers who visit the region each year.

Chapter 11

Hazards of the Indian Ocean

Every sea has its dangers. The Indian Ocean probably has a few more than most because it is situated mainly within the tropical zone. For those who dive in this area it is as well to be forewarned and, where possible, forearmed.

A black-tipped shark in Ceylon waters.
(Arthur C. Clarke: Seaphot)

The sea – as we know it – has many facets. It is at once a domain of indescribable beauty and often incredible savagery. It is also a realm of a million ironies for it treats no two creatures alike; while it nurtures some it destroys others.

But in the overall farrago of life – procreation, elimination and destruction – there is a definite pattern discernible. Even if we are not exactly sure how a particular kind of species – be it fish, shark or shell – will react when disturbed, we have come to understand enough of the sea to know where special care is needed, and where it is not. In much the same way most harmless fish are wary of us when we are in the water with them; we can get close to most of them, but not too close.

For those who leave the security of the shore behind them and descend into the realm of the underwater, it is necessary to know the sea and its inhabitants. The Indian Ocean is no different from the rest, although in the final assessment there are perhaps many more dangerous denizens in the warm waters off Africa, India and Western Australia than one would expect to find, for example, off the coast of Europe.

I will discuss only those creatures of a hazardous nature which are fairly common in the Indian Ocean. Sharks and stonefish are dealt with in more detail elsewhere in this book. The more obscure killers are left to esoteric handbooks, for they do not really concern the amateur diver.

Although some of the creatures met while diving off Moçambique, Madagascar or Malindi are hazardous, they are not necessarily dangerous all the time. To anyone really interested in fish life in the littoral zone, some of these creatures can provide a fascinating insight to the ways and wiles of many of the ocean's inhabitants; they can offer those who respect their potential a really interesting field of study.

Unusual creature in any ocean—a toby or *blaasop*. These fishes (*Amblyrhynchotes henchoni*) are poisonous to eat. (Gary Haselau)

Jellyfish: A regular question asked of most diving instructors is whether jellyfish can sting. The answer is a general yes, though not all such stings are very painful. One of the characteristics of a jellyfish is the presence of tentacles, all of which are equipped with nematocysts, or stinging cells.

Probably one of the most deadly venomous creatures in the world, ironically, is a sea wasp which is found in the Indo-Pacific between Australia and the mainland Asiatic coasts. It has been known to cause death to a swimmer in a matter of seconds and is usually fatal in minutes. This species is small, usually no more than four or five inches across the nearly transparent 'bell'. Tentacles reach a length of between twelve and fifteen feet.

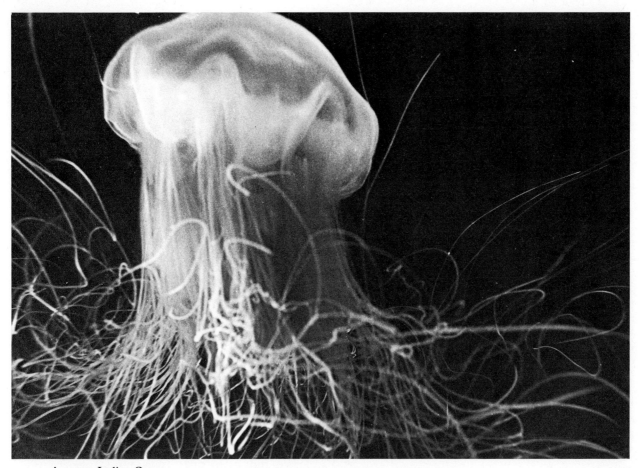

A young Indian Ocean jellyfish. These creatures are common throughout all oceans and are also found in fresh-water lakes. (Colin Doeg: Seaphot)

The largest of the jellyfish is *Cyanea capillata* (*Linnaeus*) with a huge 'bell' measuring more than ten feet across and tentacles trailing 100 feet. Severe stings from this species may produce immediate unconsciousness, frothing at the mouth and breathing difficulties.

Generally, the species found in Indian Ocean waters are not particularly dangerous. In five years of diving I have never been stung and often the sea has been thick with them. One unfortunate international-class spearfisherman who often dives off the coast of South Africa has, over the years, built up an allergy to jellyfish stings. When stung, his face blows up like a football and he can hardly see out of his eyes.

Sea Nettles: These are tiny jellyfish which are difficult to avoid as they are too small to be seen with the naked eye. They are uncomfortable rather than dangerous and will cause small painful blisters very much like nettle stings. The blisters usually disappear in an hour and there are no after-effects.

Portuguese Man-of-War (Bluebottle): Of all the many fascinating creatures of the sea the bluebottle is perhaps the one that is least understood. We tend to speak of it as an animal when in fact it is of the jellyfish family. Although the amount of sting imparted by the bluebottle is fatal to most small fish, it is rarely enough to kill a human.

The Man-of-War is a drifter at the mercy of wind and wave and for this reason is often washed up on to Indian Ocean beaches. *Treatment :* A recent discovery for the treatment of these stings is to rub ordinary kitchen meat tenderizer into the sting area. This effectively neutralizes the toxin and produces immediate relief.

Sea Anemones: Anemones are the flower-like growths commonly found in rock pools. They have thousands of tiny stinging cells on their numerous tentacles and are very common in East Africa where they can grow to a fairly large size. Fortunately the stinging cells are usually too weak to penetrate human skin and only people with very sensitive skins will be affected. If a sting does take effect it will appear as a painful rash on the skin and can be treated as you would a normal bee sting.

This fine example of co-existence between a clown fish and a deadly sea anemone is not unusual. One sees this kind of dichotomy in many spheres of underwater life.
(Peter Saw)

Coral: Divers whose underwater adventures take them into the clear warm tropical waters of the Indian Ocean know the splendour of coral formations. Some of the most spectacular sights in the undersea world are to be found around the coral reefs of East Africa, the Seychelles, Comores and Madagascar. Although not nearly on the scale of the Red Sea or Australia's Great Barrier Reef, the islands often present fascinating coral contrasts. But these beautiful growths also have their dangers.

Beautiful coral reef formations are found throughout the length of the tropical African coast. (Conway Plough)

Much of the coral around warm-water reefs are razor sharp. Some of it is poisonous. Thus divers who operate in shallow waters at low tide or when there are strong currents or rough seas often get catapulted into a reef by a sweeping wave which will give them an experience they will not soon forget.

There are a number of types of dangerous coral. The first of these is fire coral, a flat, honeycombed coral formation that grows on shallow reefs down to fifty feet or more. It is pale brown to mustard-yellow or browning-yellow depending on prevailing light conditions. There are two members of this family in the Indian Ocean: *Brown Fire Coral :* This occurs as smallish brown fern-like growths, rarely more than a foot high. It grows mainly on large coral heads and rock structures. *Yellow Fire Coral :* This forms much larger growths which can form large convoluted coral heads. They are fairly solid structures with thick yellow branches. All fire coral has a powerful sting, like touching a hot stove. Full contact may cause a burn that will fester and burn for days. If coral cuts are left untended they can become an ulcer with a septic, ploughing base within a few days.

No fresh coral should be handled with the bare hands. If you wish to keep a coral make sure it has been cleaned properly beforehand and left in the sun for an adequate period of drying out.

The most attractive coral, perhaps, is the staghorn, which together with the elkhorn makes a fantastic undersea backdrop in any collector's showcase. Because of the time it takes for coral to grow to maturity it is not recommended that live coral be moved unless absolutely necessary.

Starfish: All except one of the Indian Ocean starfish are harmless and may be handled safely. The dangerous variety is the Crown-of-thorns starfish, *Acanthaster planci*, which lives on living coral and has reached plague proportions in the Pacific where it has destroyed many coral reefs. This starfish is easily recognized by its large size,

The Crown of Thorns starfish (*Acanthaster planci*) can inflict a nasty sting. (Conway Plough)

sometimes over two feet in diameter. Its colour is red and white and it has more than a dozen arms. The whole of the upper surface of the starfish is covered with inch-long spines which are coated with a toxic mucous. Injury by the spines causes severe pain and nausea and should be treated by a doctor. The Crown-of-thorns is quite common on some of the offshore coral reefs.

The Molluscs: The molluscs comprise all the snails and sea-shells of the shore and also the octopus and squids. Of the first group there is one dangerous family, the Cones.

These animals possess a toothed device called a *radula* which can inject a very powerful venom. Several people have died from the sting of a *textile* cone and I know of one case in Zanzibar where the victim was hospitalized for several weeks. Cones are very common in all Indian Ocean waters, especially in sandy areas. Not all are dangerous to man but unless you know the group well it is best to treat them all as potentially lethal and handle with extreme care, especially the *geographus* variety which is deadly. (See Chapter 13.)

Sea Urchins: The most common sea urchin in the Indian Ocean is the black-spined urchin *Diadema*. These are easily identified as small black globes with very long black spines which are coated with a mildly poisonous slime. The spines penetrate easily and painfully and invariably break off in the flesh. The painful irritation caused by the poisonous slime usually wears off in ten to thirty minutes if the penetration is not too deep. Unless complications set in the spines can be left in the flesh till they dissolve naturally without ill effect. Trying to get the spines out with a needle or such-like device may cause complications which could result in a septic wound. Urchins are common in all coastal regions and on all coral reefs. They frequently gather in large groups.

A more dangerous urchin is the orange sea urchin, *Toxopneustes*. This is a rounded, slightly flattened urchin, about four inches across, orange-yellow in colour with short spines and what appears to be a star-like pattern over its upper surface. This pattern is made up of hundreds of pincer-like devices, each with three poison sacs which can inflict a nasty sting. These urchins are uncommon on the coast and only occasional specimens will be found on the coral reefs.

The Giant Clam: These are found on most reefs in the Indian Ocean though they seldom grow more than a foot long. They have very powerful muscles and can deliver a powerful pinch to the unwary hand or foot. There is one recorded incident of death resulting from someone having been caught in the vice-like grip of a giant clam. This happened off Nyali Beach north of Mombasa some years ago when a tourist stepped into the opening of one of these creatures. He was trapped and drowned when the tide rose over his head.

Scorpion Fish and Stonefish: There are many deadly fish in the oceans but these two tropical species are often found in shallow Indian Ocean waters. The scorpion fish can cause intense pain if touched and has led to fits of fainting in the past and needs prompt treatment. (See also Chapter 4.)

Scorpion fish, also called turkey fish, paradise fish, lion fish and many other names, is found throughout the tropical Indian Ocean. (Peter Saw)

The stonefish is even more deadly. There are many recorded instances of death following a sting by this ugly warm-water critter. The pain which follows the entrance of a barb of this fish is regarded by some as the worst imaginable. Peter Saw of Malindi, Kenya, stood on a stonefish while cleaning his boat. The barb entered his toe. Within seconds he was in a convulsion of pain which he recalls with horror to this day. Only the presence of a doctor with the necessary antidote offered any relief and in any event he was laid up for a week.

Both fish are very slow moving and passive and react only if attacked. Unfortunately the stonefish, which looks like a stone or piece of dead coral, lies immobile in the sand or in rocky crevices for most of its life and does not move if a foot or flipper approaches.

Treatment : It has recently been discovered that the best treatment for this kind of sting is the immediate immersion of the limb affected into a bucket of extremely hot water (above 55 °C). The water alone will cause burns but it will be found that the much greater pain inflicted by the stonefish poison will be instantly relieved. Hot water neutralizes the toxin. Epsom salts or even common salt can be added to the water for the beneficial hypertonic effect.

If hot water is not immediately available, irrigate the afflicted area with salt water. Clean the wound and remove any barb or sheath to encourage bleeding. The use of ice, ammonia or potassium permanganate is not recommended. *Note :* The Scorpion fish is also known as the tigerfish, zebrafish, paradise fish, turkey fish, lionfish, and by other epithets in various parts of the world. There are eight species in the Indian Ocean.

Sting Ray: These friendly creatures will usually shy away unless trodden upon by accident or taunted. They strike with a serrated barb imbedded in the tail and this action secretes a quantity of glandular poison into the victim. Neglected stabs may cause gangrene. When entering a lagoon it should be sufficient to shuffle forward on your flippers to drive away any rays lurking in the sand.

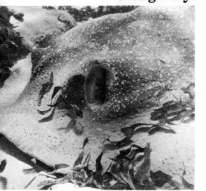

Ray. (Peter Saw)

It should be noted that not all rays sting. A fairly large family of electric rays known as *Torpedinidae* are found in most warm- and temperate-water zones and some of these, though circular and flabby in appearance are capable of developing a 200-volt shock which will stun a man. These rays frequent sandy or muddy bottoms and are usually excellent at camouflage when at rest against a dark background.

Moray Eel: These sea creatures range in size from a few inches to almost twelve feet in length and can be extremely dangerous to those who are accustomed to sticking their hands into holes in rocks or reefs. They are prevalent in warmer oceanic waters, though an equally dangerous Conger Eel (*Linngeus*) is found in colder southern latitudes and grow to eight feet. Morays have muscular jaws, powerful bodies and strong teeth, but rarely attack unless provoked;

Moray eel. (Peter Saw)

1. Madagascar people are colourful, friendly and possess an infectious *joie de vivre*. (Brian Rees)
2. Seychelles beach near Anse a la Mouche. (Peter Saw)
3. Fishermen off Praslin in the Seychelles group use nets to make their daily catches. (Peter Saw)

2 ▶

▲5

◄4

6
▼

they can inflict a nasty bite. Best left well alone.

Although Morays have often been accused of inflicting a venomous bite there is no evidence for this. Part of the reason for this argument is that unless promptly treated a Moray bite can turn septic within hours. Wounds are usually jagged and should receive the same attention as shark bites.

Octopus: No diving book would be complete without a comment on the cephalapod which is found throughout the Indian Ocean. Most of the octopi you will come into contact with will be found to be shy, reticent creatures, more anxious to disappear into the haze than befriend you. Rarely will you find a specimen larger than about three feet across. In other oceans this creature can grow to about twenty feet across. The beak of a larger octopus can inflict a severe bite if not carefully handled; to be admired but not touched.

Divers in Seychelles waters are likely to encounter Barracuda from time to time – some up to lengths of six feet. Although these predators appear aggressive – their movements are fast and erratic – the best counter is strong, even movements in their direction and they will give way. Avoid taunting, as although 'cuda attacks are rare, they are not unknown. Fortunately the Indian Ocean family (*Sphyraena japonica*) is less aggressive than the variety found in the Caribbean.

Sea Snakes: While eels are not venomous, there are about fifty species of true snakes that have returned to life in the sea. Most live in shallow waters about Asia or Australia but one, *Pelamis Platurus* which rarely grows to more than three feet in length, is pelagic and occurs throughout the Indo–Pacific zone. Easily recognized by the flat paddle-like tail boldly barred in orange and black, the body is sharply zoned – dark dorsally and yellowish or dirty-white vent-

---

4. A speckled rock cod views the world warily from the entrance to its coral hideout. (Johnny van der Walt)
5. The diver explores among the rocks. (Walt Deas)
6. Feather star in its fully exposed position. If disturbed this creature withdraws into its elongated habitat. (Peter le Lievre)

rally. Beautifully marked as they are, sea snakes are to be treated with the greatest respect for every single one has fangs and poison, some more deadly than the cobra. Fortunately they are only aggressive when provoked. The bite of a sea snake shows no symptoms for twenty minutes to an hour. Aching and stiffness gradually pass into paralysis from the legs up. An estimated 25 percent of the victims die.

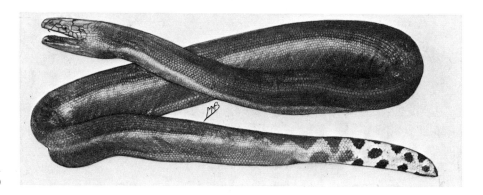

Sea snake (*pelamis platurus*). (J. L. B. Smith)

Although it is highly unlikely that any casual diver will be attacked by a sea snake, it is as well to know what to do in such an eventuality. Do not cut or suck the wound. Avoid any exertion and a tourniquet should be used. The nearest doctor should be contacted.

Chapter 12

Descent off a Moçambique Reef

Some of the most isolated diving spots in the world are to be found off Portuguese East Africa. Undersea photographer **Johnny van der Walt** *spends most of his holidays off this tropical coastline which he regards as one of the most exciting to be found anywhere. Apart from a multifarious and colourful marine life, says van der Walt, there are wrecks, sharks and numerous reefs which have not yet been dived on by man.*

I looked up from a delicately hued snapper fish which I was trying to photograph to see my buddy beckoning frantically. As I swam towards him he gave the underwater signal to indicate the presence of a shark, and then gestured that I should follow him. Intrigued, I did so, and as he led the way my mind played over a dozen possibilities.

We glided still farther downwards through the clear warm sea to a deep overhang in the reef which I had cursorily inspected a short while before and had passed by. Momentarily my friend stopped at a large opening in the sheer coral face and pointed at a patch of white at the far end of the cavity. I eased myself slowly under a deep ledge along the sandy bottom at about ninety feet and as my eyes became accustomed to the gloom I realized with a start that I had come face to face with what appeared to be a sleeping shark about six feet in length.

Unquestionably the shark was alive; I could see normal gill movements and an occasional tremor of its tail. I was also uncomfortably aware that according to the books I had read sharks did not sleep; not this variety at any rate, for it was Zambezi shark, *Carcharinus leucas*. (See Chapter 8.)

But here was I with a fair-sized shark apparently oblivious of my presence and rather ominously within touching distance. Not to be outdone of the rather interesting prospect of being the first person to photograph a sleeping 'Zambe' I inched forward to take my pictures.

Each flash from the camera unit caused some anxious moments. I was vividly aware that my body was blocking the only exit from the fissure and should the shark suddenly awake from his snooze I would be in a difficult position. However, by the time I had exhausted my supply of flash bulbs it had still not stirred so I left as carefully as I had arrived, realizing that I had witnessed a most unusual phenomenon.

This was one of many interesting dives experienced on various expeditions to Moçambique. On this occasion we were on a three-week visit to Ponta Barra Falsa, a long stretch of coral reef about 400 miles north of Lourenco Marques, Moçambique's colourful capital city. The largely unexplored reef we were on stretched northwards for almost 100 miles towards Vilanculos and the island of Bazaruto, famed in international game-fishing circles for the size of its fighting black marlin (the biggest of the game fish), barracuda, sailfish and shark. Bazaruto has a number of world records to its credit.

Spearfishing on the Ponto Barra reef is excellent. The reef abounds in fish such as grouper, niggerlips, bream, tasselfish, spadefish and parrots of all shapes and colours. Game fish tend to hang over the edges and gaps of the reef in groups. Large numbers of queenfish, kingfish, barracuda, sea pike, bonito are regularly taken. The area is well known for a very large variety of kingfish which usually swim about in pairs and show little regard for divers. When shot they are regarded as 'dirty' fighters running near the bottom and around boulders and rocks. Heavy line is needed on one's speargun reel to avoid being smashed up.

Another favourite spot for spearos is at Ponto Tofa, the main beach of Inhambane. The reef is situated at the point of the bay and consists of a large slab of rock dipping from the point to the north and east. The inner edge is comprised of large boulders.

On the outside edge there are 'cuda and sea pike in profusion. When feeding, these fish will swim over the reef and into the bay where they tend to take an active interest in any diver in the water. A spearfisherman can stalk these fish by swimming slowly towards them from behind and below. When approached in this manner a barracuda will usually turn broadside on, so presenting an easy shot.

Other interesting dive spots in this area include Cabo de Inhambane, Jangamo, Inharrhugo, Cabos dos Correntes with its distinctive

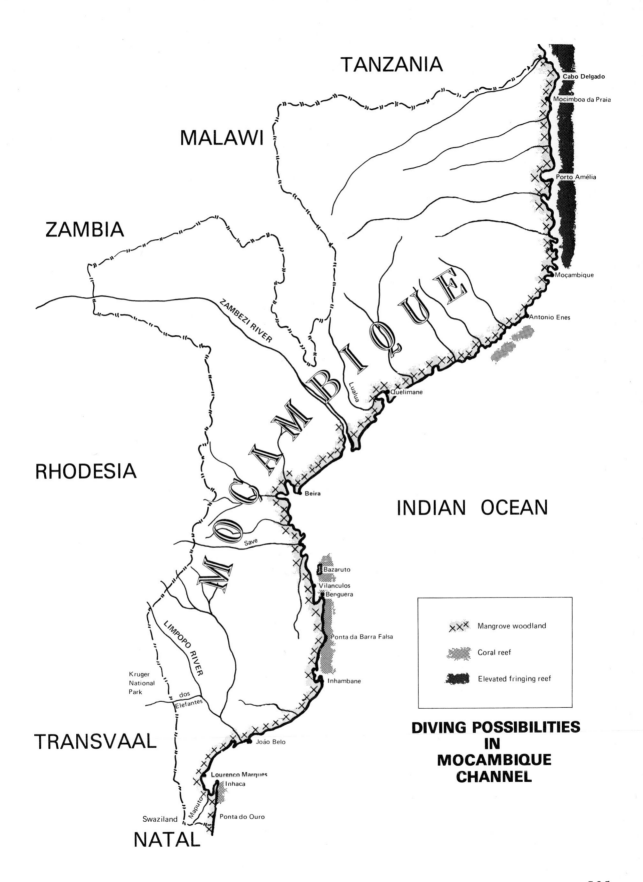

TANZANIA

MALAWI

ZAMBIA

RHODESIA

TRANSVAAL

NATAL

MOÇAMBIQUE

ZAMBEZI RIVER

LIMPOPO RIVER

Kruger
National
Park

dos
Elefantes

Swaziland

Maputo

Lualua

Save

INDIAN OCEAN

Cabo Delgado
Mocimboa da Praia

Porto Amélia

Moçambique

Antonio Enes

Quelimane

Beira

Bazaruto
Vilanculos
Benguera

Ponta da Barra Falsa

Inhambane

João Belo

Lourenco Marques
Inhaca

Ponta do Ouro

	Mangrove woodland
	Coral reef
	Elevated fringing reef

**DIVING POSSIBILITIES
IN
MOCAMBIQUE
CHANNEL**

135

The coral in Moçambique waters is excellent. (Johnny van der Walt)

lighthouse, Big Rock and Masinga, the latter an area presenting those interested with some fascinating wrecks to explore. Many divers regard the Sylvia Shoal as one of the finest to be found along the entire Moçambique coast. This area is excellent for hunting sharks with explosive-tipped spears.

But to really get the feel of diving in Moçambique waters it is necessary to start at the southern tip of this Portuguese colony which stretches northwards towards the Tanzanian border in a long meandering line for the best part of 2 000 miles and no place provides a better introduction to the Indian Ocean than Inhaca, which lays claim to the rare distinction of having the most southerly coral reef in the world. Because the island lies within an area of unusual patterns of current eddies, Inhaca is regarded by many as a naturalist's – and a diver's – paradise.

The marine fauna and flora at Inhaca are a mixture of tropical forms seeded by the warm Moçambique current and temperate variations extending northwards from Natal and the Eastern Cape. The island consequently lies in a region of transition from warm temperate to tropical and among other things is regarded as ideal shark country. This is the traditional home of the Indian Ocean Zambezi shark.

For a diver, the inner shoreline facing the bay is sandy with a few rocky points. The *milieu* is fringed by large areas of flat, muddy sand covered by sea grass. Aided by the effluent of nearby river estuaries Inhaca has become an ideal breeding ground for a variety of pelagic food fishes. For the novice and younger set diving here is ideal as sharks rarely enter the inner reef area.

Although relatively small – Inhaca measures only about seven miles by four – diving possibilities are multiple. Coral reefs on the lee fringe are sheltered and shallow and it is possible to float for hours over formations which extend in places into fairly deep water.

High and dry on the sands of a Moçambique beach lies the *Berea*, one of many shipwrecks in these waters. (Johnny van der Walt)

On the seaward side the topography is typical of much of the coast-line from Ponta do Ouro in the south to Inhambane farther north; wide sandy beaches sweep up to clusters of rocky outcrops many of which form northward-pointing reefs running out from the shore. Here, apart from the occasional wreck (and the Moçambique shoreline is notoriously hazardous as far as shipping is concerned) there are enough varieties of sponges and bryozoans to provide an incomparable wealth of colour and diversity. What is also interesting to those looking for additional excitement is that depths range from inshore shallows to about eighty feet and deeper but the ocean floor is always broken by numerous gullies, overhangs and precipitous drop-offs. What is more is that most of these reefs are close to the shore and many divers are able to swim off from the shore without having to battle against cross currents on the way in.

Lying about six miles offshore from Inhaca is the Baixo Danae tropical shoal with an average depth of about forty feet. Surrounding waters plunge down to 200 feet or more but the shallower area is renowned for its sea life and unspoilt reefs. Because the area is only accessible to those with boats Baixo Danae is likely to retain this primeval quality although again, divers should be cautious of entering the water alone because of sharks.

One of the surprising features of diving in Moçambique is the excellent grid of tarred roads leading northwards out of Lourenco Marques. Although these deteriorate markedly the moment one branches off for a particular beach or reef, it is possible to travel more than a thousand miles in an ordinary saloon car with passengers and heavy diving gear. A number of local excursions always include their own compressors as little provision has been made for this kind of facility by the Portuguese authorities. Divers are also advised to keep a strict check on depth limitations as only one decompression chamber exists in the entire country and that is at Lourenco Marques in the south.

After diving in murky Southern African or European waters the clarity of the Benguera area is an experience. The first time I dived off the northern Baixo Zambia shoal the prevailing southeasterly wind had dropped and long before we had reached our target position we could spot miles of coral reef beneath the boat as we sped across the water, often as much as fifty feet below. I only fully appreciated how clear the water was when, sitting on the bottom of the sea near the end of a 150-foot anchor chain I could clearly recognize the divers climbing out of the boat into the water above me.

It's the islands farther north that provide a real touch of the exotic. Varying in size from a few hundred yards across to a few miles, these scattered strips of beach and coral which are a continuation of the Vilanculos Peninsula are fringed by acres of coral reef. Apart from the usual rocky outcrops liberally sprinkled with coral one also finds large shoals of coral making the area truly tropical when compared to conditions farther south. Unfortunately the region is being extensively developed for the tourist market and this influence is already being felt as it is on the Kenya coast.

Although coral appears at places all along the length of this stretch of coastline it does not generally take the form of a solid barrier reef. Patches of coral grow larger the farther one travels north as do the number of species of fish. There is, however, a notable lack of coral from the Save River south of Beira northwards nearly as far as Antonio Enes. However, at this point one encounters the first

Cave and blow hole near Ponta Barra Falsa. (Johnny van der Walt)

real elevated coral fringing reef, which continues northwards to the frontier and into Tanzania and beyond. The presence of this reef is indicated by numerous small islands which dot the coastline and lie about twenty to thirty miles offshore. This expanse of coast remains largely unexplored; although several expeditions have made it to Porto Amelia, Nacala and farther north they have barely covered a fraction of the terrain. A large section of this northern territory remains inaccessible by road and such tracks as exist are menaced by periodic cyclones, floods and other hazards. When visiting the area, as in the region of Ponto da Barra Falsa, one also has to take into consideration that this is a section of Africa where there is still much wild game; wandering lions and herds of elephant have proved a problem to camping divers in the past. But if it's the unusual you are after you have come to the right place, both in and out of the water.

For the occasional visitor, the Moçambique shoreline can be divided into various conditions or zones. The first of these is the typical rocky point with associated reef.

Typical rocky point with associated reef

A. (As found between Ponta do Ouro and Inhaca)

This coastline is characterized by a series of gently curved bays punctuated by outcrops of rock which protrude from the sandy shore and run at a tangent of about 15°–30° out from the beach. The north-flowing counter-current sweeps loosely around this natural breakwater and scours a small bay into the generally straight coastline. The sand south of the point is coarse and forms a steep slope to the breakers. Around into the bay one finds a wide flat beach composed of fine grains of sand. This sand is not the white coral debris so common farther north but the siliceous type found farther south.

The reef usually starts on the beach at about the high-water springs level and gradually sinks as it runs out to sea. It usually continues underwater for about half to one-and-a-half miles before sinking beneath the ever-present sand once more. Closer inshore small flat patches of rock are usually to be found in the centre of the bay just beyond the surfline. Quite often parallel lines of reef run farther out to sea in deeper water and are not visible from the beach.

The inter-tidal habitat is typical of the warm temporal regions of Zululand and Natal sprinkled with quite a few tropical intruders.

Typical rocky point in
Moçambique; barnacles
covering high rocks leading
down to algae in rock pools.
(Johnny van der Walt)

The rocks are covered with periwinkles, *Littorina obesa*, barnacles
of various types and oysters *Crassostraca cucullata*. Near the water's
edge the rocks are hidden by numerous types of algae which provide
cover for a variety of creatures including limpets, worms, crabs and
so on. Large colonies of zooanthid anemones are common as are

sponges and turnicates. Rocky pools provide a haven for sea cucumbers, starfish, sea urchins, cowries and cone shells. Tucked away in algal tufts in the rock pools are incredibly beautiful nudibranches. Fish of all varieties are also found in these pools. These include *Chaetodontids*, *Abudefduf* species, blennies, gobies and snappers, as well as the juveniles of many deep sea fish.

As the reef proceeds underwater the ecology alters but the basic pattern remains. Coral is found growing in sheltered spots and a surprising number of species are represented. The *Octocorallia* or 'soft' corals are plentiful and varied while isolated branches of the true or 'hard' corals peer out of crevices and potholes along the length of the reef. For example the Research Station on Inhaca has collected and identified 135 different species of coral comprising forty-four genera from the two coral reefs associated with the island.

The fish that are found in the warm waters show a mixture of two adjacent regions. Inhaca forms a 'gateway' to the tropical region of East Africa while retaining the temperate fauna of Southern Africa. There are a few fishes of the tropical Indo-Pacific which are not to be found at Inhaca but only appear further North. Among the invaders from the South may be found the musselcracker, *Cymatoceps nasutus*, and the kob, *Johnius hololepidotus*. Indeed kob have been shot as far north as Inhambane. The large gamefish such as marlin, tunny, barracuda and kingfish abound in the open sea and provide ready sport for spearfishermen. (It is illegal to shoot fish when scuba diving in Moçambique waters). One fish of special interest in this area is the stonefish, *Synanceja verrucosa*. Inhaca is the farthest point south that this variety is found. All the tropical sharks are found along this coastline and have been regularly seen by divers.

B. (As found between Chonguene and Inhambane)

Large sand dunes sweep down to essentially the same sandy coastline described earlier. The rocky points are, however, noticeably larger and usually stand about three-feet above high-water springs. These then form a barrier-type reef which allows a generous growth of coral in the sheltered enclave. Fauna and flora in this area is more decidedly tropical with fewer invaders from the south. Various parrot fish, wrasse, squirrel fish and other coral beauties normally found around the Seychelles make this area their home. The water temperature rises steadily as one proceeds north and this enhances the already magnificent diving.

Typical offshore shoal

This takes the form of a rocky reef rising from a flat hard sand bottom at about 200 feet to near the surface. The shoal lies about four miles offshore and is commonly a mile or two long by half-a-mile wide. The shoal is far enough from shore to lie in the edge of the clear Moçambique Current which also ensures that the format of the reef is decidedly East African tropical. Coral covers most rocks and provide a home for countless tropical animals associated with true coral fringing reefs. Add to this the presence of many large gamefish brought in by the deep sea currents together with steep drop-offs on the seaward side to 300 feet plus and you have a miniature, self-contained, divers' paradise. Sailfish, dorado and wahoo are included in the ranks of the gamefish and huge manta rays are a common sight over the reef. The large whale shark, *Rhincodon typus*, can often be found, particularly near the edge of a plankton bloom. Certain areas seem to be regularly visited by these lazy giants suggesting a configuration of currents leading to regular deep water upwellings and consequent plankton growth. The groupers increase in size and species and can number in their family some of the prettiest (and tastiest fish) in the ocean. However, it is only far north at Porto Amelia where certain species are reported to attain weights between 1 000 to 2 000 lbs.

Fringing coral reef

Representative of the typical East African tropical area, this reef starts somewhere near Antonio Enes and continues right past the border up into Tanzania and Kenya. The beaches are composed of the white sand resulting from coral breakdown and the reef is solid coral, the living building upon the dead. A profusion of various *echinoderms* is characteristic of a coral reef. Every nook and cranny hides multitudes of brittle stars. Sea urchins hollow out tunnels along the length of the reef and brightly coloured starfish are common. Crown-of-thorns, *Acanthaster planci*, occurs but no reports of any epidemic have reached South Africa. Large tracts of this region are never dived on and consequently serious infestations could quite easily be missed.

The molluscs are strongly represented as a group of which the best known is perhaps the giant clam, *Tridacna*. Many types of cowries, cone shells, turbans, nudibranches etc., are found as well as many different bivalve molluscs.

The crustaceans form a large and multi-coloured group living on or near the coral reef. Several tropical species of lobsters are represented, these belonging largely to the *Panulirus* genus. Shrimps come in all sizes, shapes and colours and live in the most unusual

Crown of thorns starfish.
(Conway Plough)

places. Quite often shrimps can be found in molluscs, on anemones, nudibranches and, surprisingly in the gill chambers of fish.

Finally, the most obvious and spectacular denizens are the fish themselves. These are the typical tropical Indo–Pacific types found round the Seychelles, Madagascar, Tanzania and Kenya. The very names conjure up visions of colour, shape and size which only barely approach the startling reality. Butterflies, moorish idols, *demoiselles*, clown fish, rainbow fish are all names which reflect some of the beautiful bold colours that nature uses so lavishly.

Chapter 13

Heritage of the Indian Ocean

Internationally-renowned conchologist **Lallie Lee Didham** *who has spent most of her life along the East African coast tells of some of the more exotic sea-shells to be found in the Indian Ocean.*

In all its many forms, conchology – the pastime (some say art) of collecting sea-shells – is probably as old as man himself. From the beginning of time the oceans of the world have yielded up a vast and varied assortment of sea-shells ranging in appearance from downright ugly to breathtakingly beautiful, and the rulers of the land have been gathering them as assiduously as the sea has been delivering them.

These days serious shell-collecting is largely practised by specialists, scientists and enthusiastic amateurs, but in the earlier period it was a labour of necessity and profit. Sea-shells have played an enormously important role in the economic history of the world. For centuries the long East African coastline from the Red Sea downwards supplied Indian Ocean shells to all of Africa, and many of the cowries used to decorate the shrines and figurines in the ancient Nigerian cities of Benin and Ife originated 3 000 miles away in Sudan and Somalia. They came to West Africa by sea and across the Sahara by camel becoming a valuable form of currency which found a ready demand in the markets of Kano and Timbuktu besides the populous coastal tropical centres farther south which we know today as Takoradi and the Niger Delta.

Early man collected shells for food, for adornment and for prestige. Certain shells were used in magical ceremonies; others were worn as good luck charms; cowries in particular were sought after to ensure fertility, as their shape resembled a reproductive organ. Others were reserved for the use of kings, and even today the golden cowrie of Fiji and the New Hebrides is only worn by chiefs.

The little yellow money cowrie was once used as currency in many parts of Africa. Threaded on strings, forty at a time, tons of them were transported to West Africa to pay for gold, ivory, palm oil and slaves. Twenty to fifty thousand would be needed for a slave; sixty to a hundred thousand for a beautiful young wife. One man built a bungalow and paid for it entirely with money cowries; a staggering sixteen million of them.

Shells had other uses too. The Phoenicians founded a most profitable dyeing industry based on the little murex shell from which they extracted a purple colour which was as fast as the most sophisticated dyes used in industry today. With the exception of precious stones and rare metals, the most expensive commodity of the ancient world was probably a dyestuff made by the Phoenicians from a fluid secreted within the shells of three marine molluscs, the Banded Dye Murex, the Spiny Dye Murex, and the Rock Shell. This dye was called 'Tyrian purple'. Around 300 A.D., during an inflationary period, a pound of silk dyed in Tyrian purple cost an amount equivalent in purchasing power to about £10 000. Or, to put it another way, a typical silken scarf worn by a woman today, would have cost the equivalent of £400. In the 8th Century the British used a shell for dyeing parchment; the little purple or 'dog' periwinkle. The Aztecs too used shells for colouring their artifacts. Instead of milking the snails, a custom still practised by the Mixtec Indians of Central America, the Aztecs, like the Phoenicians, either removed the gland or broke open the shell, a wasteful practice that decimated the supply. Fragments from the crushed shells were piled up outside the city in great heaps and on days when the wind blew onshore, the busy streets of Tyre stank heavily.

7. Scorpion fish. (Wolf Avni)
8. An octopus-eye-view of the Moçambique coast. (Johnny van der Walt)
9. The octopus is a beautiful and often graceful creature in its own environment, and is usually content to give the intruder a wide berth. (Johnny van der Walt)
10. Bubble shell carrying an egg ribbon off Inhaca. (Johnny van der Walt)
11. The tiger flatworm contrasts brilliantly with its otherwise drab surroundings on a Moçambique reef. (Johnny van der Walt)
12. A candy nudibranch alongside a soft coral growth in Moçambique waters. (Johnny van der Walt)
13. The beautiful black mantle of an *Ovula Ovrum* shell was found in the Indian Ocean. (Peter Saw)

▲7

8▶

9▼

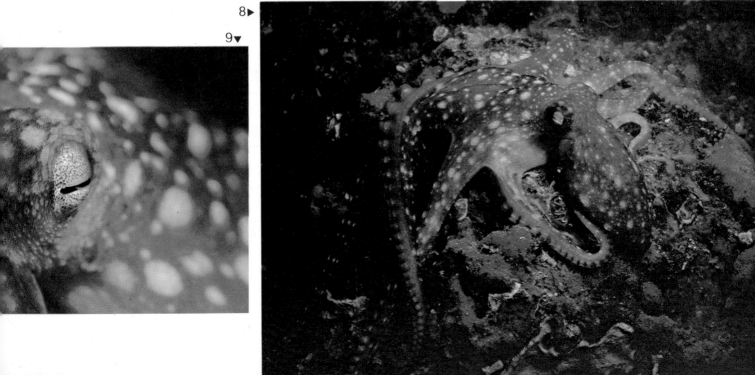

▲10

11▲ ▼12

14▶

13
▼

15 17 16

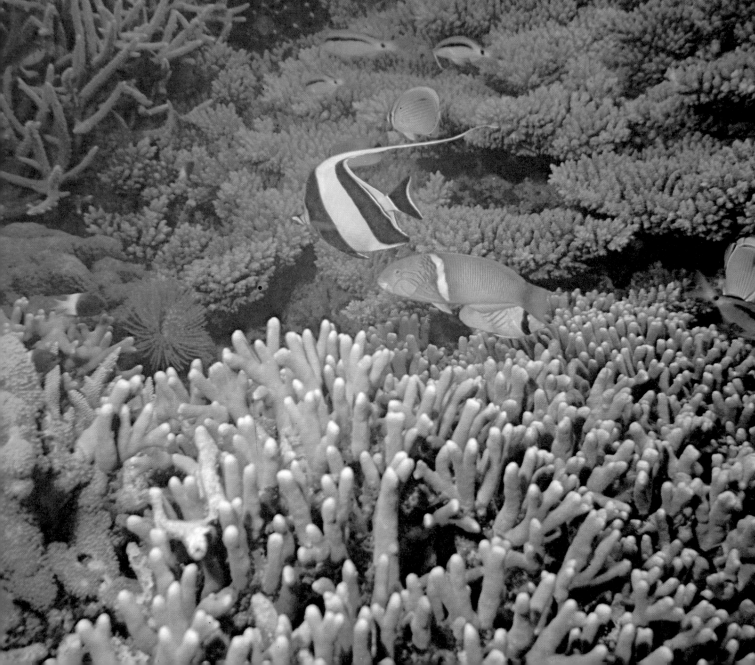

Shells have provided man with cloth. The pinna shell, a long slim dark bivalve, attaches itself to a solid base or rock in muddy sand that it frequents by a *byssus*, a silky thread that it spins itself. The Romans wove cloth from it, and in the Middle Ages the pinna's anchor-cord produced the famous Cloth of Gold. Until quite recently, there were still gloves and capes woven from the byssus thread in Sicily although it was discreetly adulterated with a proportion of one-third silk. The Italians also used tiger cowries to iron their lace and burnish their paper.

Shells have always made perfect utensils and tools from the earliest times. We use them still: Some fish dishes are delightfully presented when served in shells and I use some shells for the arrangement of flowers and others for bird-baths. Before modern society came to rely on plastics, most buttons were made from the shell of the pearly oyster and the *Turbo* and *Trochus* shells. The ubiquitous cowries were used as ballot stones by the Greeks and Romans and to this day they make ideal counters for games.

Shells are ideal for decoration as well. In the 18th Century entire rooms were encrusted with shells and at one stage there was a vogue for grottoes, which were considered incomplete without them. Shells were used on snuff and other boxes, on frames for pictures and mirrors and as goblets. They were also used as jewellery, and still are, either on their own or as pieces cut from the pearly nacreous inner layers. Cameos are fashioned from shells, each layer displaying a different hue. The best cameos are made from helmets, *Cassis rufa* and *Cassis cornuta*; but in the Western Hemisphere they use the big conch, *Strombus gigas*.

A shell with a different role indeed is the big triton conch, *Charonia tritonis*, which is sometimes a trumpet. It produces a splendid resonant note and was used for important ceremonies and on the

---

14. A Clownfish against an anemone backdrop in Seychelles waters. (Peter Saw)
15. The Lion fish preens itself in Mauritian waters. (Peter le Lievre)
16. In spite of a granite shore area, the Seychelles islands have nurtured a variety of coral species. (Peter Saw)
17. The Malindi area of Kenya provides some of the most beautiful coral beds to be found in any ocean. (Peter Saw)

battlefield by the Polynesians. In East Africa triton shells were used in the past to urge ferry hands to greater efforts when cars were pulled across Kilifi creek north of Mombasa by groups of labourers hauling the ferry along a chain, and when dhows enter Lamu harbour they still announce their arrival with a blast on their shell trumpets, as they have done for centuries. Indians also use their sacred chank shell, *Turbinella pyrum*, as a trumpet.

A diver examines a fine example of an Indian Ocean triton shell. (Peter Saw)

There is a great deal of romance about shells. They speak mutely of the remote regions in which they originated and they charm the soul with their infinite variety. In the past they were an additional source of wonder; great collections were formed by rich and knowledge-able amateurs who sailed forgotten parts of the seas or had agents buying from those who did.

Today most of the famous collections are in the hands of the great museums, to be enjoyed and studied by all. Yet even the savants are puzzled by some of their rarest specimens; one-of-a-kind enigmas hailing from heaven-knows-where, unique and irreplaceable.

Once I held in my hands a *Cypraea leucodon* cowrie, one of only two in the world. There are other cowries almost as rare: the *Cypraea broderipi*, for one, of which only three are known to exist and the *Cypraea valentia*, of which there are just five. No one knows their origin; they are probably deep-water shells. I have also handled the world-record golden cowrie, *Cypraea aurantium*, and the famous cone shell, *Gloria-maris*, and I have had the fun of confounding the experts. No one was able to identify my unknown *Cymatium*; it was not illustrated in any of their books, so I left it with the specialists for identification.

Exciting as these rare shells are, the greatest thrill for any conchologist is finding his own shells, whether they are rare or not. I can remember how my first white or egg cowrie looked, the primary colour almost hidden by its jet black mantle and contrasting orange *papillae* (white cowries, *ovula ovum*, are allied cowries, because they have teeth only on one side). Within months of my first find I settled down to the serious business of trying to amass the full range of different Indian Ocean cowries I could find, with every new species an exhilarating discovery. Records show that there are fifty-two different cowrie species in East Africa; I have forty-five, and the last seven are certain to present problems, but I have not given up.

Common tiger cowrie in its natural surroundings.

Most beginners are usually first drawn to cowries. The reason could be that these are often the first shells one finds in the Indian Ocean waters; the tiny *Cypraea annulus* or ring cowrie and the big *Cypraea tigris* or leopard cowrie. Their shells are radiant and glossy and need little attention once the shell fish has been removed. But to capture all the fantasy of a fine collection one should see the living cowries in their natural environment; some are more beautiful than the shells they create. *Cypraea nucleus*, the nucleus cowrie, is a small, knobbly beige shell; the fish inhabiting it has a strawberry-pink mantle with pale-green *papillae*.

Cowries lay eggs in the form of capsules, little seed-like objects gummed together, usually under rocks. The mother broods over them as a hen does, puffing out her mantle to cover them the same way that a hen fluffs out her feathers. The egg capsules vary in colour according to the species, but they also vary within the range of the same species; the reason is obscure but it could be because of the developing eggs inside. I have seen the leopard cowrie, *Cypraea tigris*, on magenta egg capsules and on greyish-mauve ones. Both the false swallow, *Cypraea kieneri*, and the Chinese cowrie, *Cypraea chinensis*, run the gamut from golden brown to pink egg capsules. The lynx varies from cream to lavender; so does the star cowrie, *Cypraea helvola*.

It is important to put rocks and weed back after looking for shells

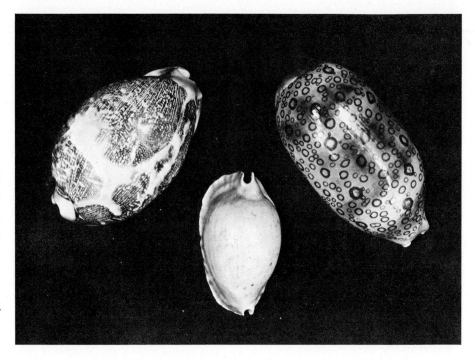

Fine examples of Indian Ocean cowrie shells. On the left is the fairly valuable pink mappa cowrie (*Cypraea mappa*). In the middle is a valuable cowrie found near Perth, Western Australia known as *Cypraea marginata* and on the right is the more common though equally beautiful Argus Cowrie. (From the collections of Al Venter and Gordon Verhoef)

so as not to disturb any eggs or other marine life that may be adhering to them, or sheltering beneath them.

Having entered the sea one soon discovers that there are many interesting other shells waiting to be uncovered. As your cowrie score mounts, and the remaining species become more and more difficult to find, most conchologists enlarge their collections to include other shells. The handsome cameo shell, the *Cassis rufa*, with its orange mouth and pink-and-brown flecked back is one of the first most are drawn to. Its larger relative, the helmet shell, *Cassis cornuta*, is more difficult to find as it tends to live in deeper water with a sandy bottom – for both shells tend to bury themselves in the sand – so I kept watch for a helmet. At last I was rewarded by a beauty. This specimen was resting on the bottom in about eighteen feet of water. Maia Hemphill from· Shimoni, a well-known authority on Indian Ocean shells was lucky enough to find one sitting on a sandbank; it was encrusted with weeds so at first she thought it was a rock itself. Not until she turned it over did she discover the true identity.

After cowries I think the favourite shells are cones. They are interesting and varied; their colours never seem to fade as some of the cowries' colours do. But they need more work, having to be cleaned outside as well as inside. While cowries are peaceful browsers, grazing on sponges and coral animals, cones are more active killers,

Cone shells taken from
Seychelles and East African
waters.

stalking their prey which include other shells, worms and fish. Fed
by a poison gland their barbed tongues can penetrate the human
skin easily, and five cone genera are known to cause bad stings.
These are the *aulicus*, *textile*, *marmoreus*, *tulipa* and *geographus*. The
last of these is a killer; it should be handled with respect and care,
being held from the back and top of the shell, away from its rear
aperture.

The spiky murex shells have a great appeal, as have the lesser frilled

The delightful Indian Ocean
Murex triremis. (Wolf Avni)

ones. This species usually lives in rather calm sheltered waters with a muddy bottom where they prey upon bivalves and other shells. The big triangular *Murex ramosus* with its red aperture and spiky exterior is a handsome variety. I have often found it with a bivalve shell clasped under its aperture, intent on opening the valves to reach the fleshy interior.

This *Murex monoon* is found in many parts of the Indian Ocean but is not all that common. (Wolf Avni)

Harp shells are other handsome carnivores. Their huge pink and white flecked feet are very obvious crawling about sand and weed. Several times I have picked one up to admire its beauty, ribbed with pink, brown and white, with an intricate inlay of the same colours between the ribs, only to be struck by the excessive sliminess of the foot. On closer examination, I have found crabs in the centre, in the process of being digested.

Hapa shell.

I have seen the big trumpet shell, *Charonia tritonis*, devouring a starfish, *Culcita schmideliana*, also known as the pincushion of the sea. By the time I arrived a third of the starfish was gone; the part next to the shell was white and rough looking. This shell's role in preying on the Crown-of-thorns starfish, *Acanthaster planci*, a tremendous menace in any coral area, is well known. You find both on live coral in fairly shallow water.

The lamp shell, *Bursa rufa*, also eats starfish. It is often found in the process of devouring the common red-and-grey starfish, *Protoreaster lincki*.

A large, beautiful and common family of shells is the *Strombidae* containing the finger or scorpion, conch shells. They are mainly herbivores, and prefer areas near sand and weed. This species often carry cloaks of weed on their backs which makes for a good disguise. *Strombidae* are very active; if you hold one in your hand, they constantly reach out with their horny operculum in a bid to turn over and escape. Their shape is most attractive. Many collectors prefer to collect a sequence of these shells, ranging from juveniles to the most mature, showing how their 'fingers' are produced from the first simple whorl, through the scale to maturity and, finally, how the 'fingers' are worn down by movement to stubs.

Sand shelling can be most interesting, following humps and bumps and tracks in the sand, preferably at night when it is cool, or when the tide is retreating, or just returning. It is at this time that most shells are active. It is possible to find bivalves, and the long pointed augers, the *terebra* shells, who prey on them, also the olives, rather like long slim cowries with a toothless aperture. In addition there are *mitres*, another handsome family of large and small carnivores, often ribbed.

Indian Ocean shells are found just about everywhere in the ocean, from the most exposed rock faces to the grimiest mud beneath mangroves. Some of the rarest varieties are found in the most unexpected places. By day, most try to escape the sun by hiding in nooks and crevices, in and under rocks, so you have to turn rocks over to reach them. By night they wander in the open. It is a sight to see a mature cowrie with its glistening mantle spread out like a veil of sheen and colour.

The best hunting ground in East Africa is in the inter-tidal area. Most specialists prefer to collect shells in eighteen inches to two feet of water, chiefly because the rocks that have to be moved weigh less in water. Also hunting when in a silty area one tends to kick up less mud by flipping about than by treading on the bottom. Another boon to snorkellers is that it is possible to stir up sand and expose shells which are often buried beneath.

Some shells do not belong to the shore at all. They are truly pelagic, coming in from the open sea. *Janthina*, the fragile purple snail, makes itself a raft of bubbles, trapping air in a slime it excretes and

An extremely rare Indian Ocean collector shell (*Xenophora*) dredged up from a great depth off Agulhas. This shell collects other shells as it moves across the ocean floor. Stuck to this one is a very rare cone (*Conus cucoronatus*) (left) and a rare unnamed volute. (From the Cohen collection Fish Hoek.) (Wolf Avni)

floats free in the ocean. After a storm, one can often gather a harvest of these shells washed ashore.

The pearly nautilus comes from Malaya and the East Indies where its octopus-like occupant hunts crabs and lobsters on the ocean floor in deep water. When it dies, its gas-filled shell floats ashore on the ocean currents.

There are many curiosities in the realm of conchology. The tusk shells, or *Dentalia* are buried in the mud like miniature elephant tusks open at both ends, some ridged and others smooth. There is also the bivalve, *Penicillus africanus*, the watering pot; it starts life conventionally, like a tiny butterfly shell, then gradually it builds a long calcareous tube with a little frill around one end speckled with holes, like a garden watering can. The mystery is how it evolves from the 'butterfly' to the 'watering pot' phase, as no one has ever seen the interim stages.

I have only mentioned bivalves as a food of other shells; but they are worth collecting, not only as a food, but in their own right as beautiful and decorative objects. This species varies enormously from the delicate rose-tinted *Tellins*, the sunset shells, to heavy ribbed *Tridacna* clams. The giant clam, often two feet across, disappeared long before the age of man. A Kenya family, the Jessops of Shimoni have a number of these fossilized giant clams in their gardens, set in coral.

Kenya's Dick Jessop found the rarest shell in Kenya, the volute, *lyraformis*, alive on a sandbank. Most other surviving specimens have been washed ashore dead, or brought in by a hermit crab. A number of deep water shells are brought in this way.

Nowadays it is more difficult to establish a good shell collection as the reefs have been overshelled and denuded of everything that can be carried away, often quite indiscriminately. Here collectors readily remonstrate with people for collecting fragile, immature shells that are certain to be broken before they reach home. It is also unnecessary to take old worn shells that are invariably discarded once they reach the smelly stage; these would be good breeding stock if left alone.

For those interested in cleaning their own shells there are a number of basic rules. Timing is vital; shellfish should be dead but not in a state of putrefaction. Big shells should be placed under a shady tree for a day or two, and the contents vigorously shaken out into the sea, as food for crabs or eels. Another effective method is to place shells in a bucket of clean sea sand, aperture downwards. They should be shaken afterwards in a basin, and then put under a fast tap or a hose to clear the fleshy portions out.

If there is still a smell, up-end the shell with water covering the source of the smell, but not so high as to show and perhaps discolour the interior. Rinse out the water every day or two and replace with fresh water until the flesh has been shaken out with the water.

The big lambis finger shell can be cleaned by securing it. The outraged occupant, in trying to get free over-extends itself to the point of no return. It is necessary to hang the lambis in a freely swinging position about eighteen inches off the ground. Eventually the fleshy occupant pulls free, or breaks off leaving only a small portion inside to remove with bent wire, or rinse out with water.

Others find deep freezing works very well. The shell muscle contracts with cold and the shellfish is easy to remove when you thaw it quickly under warm running water. I once succeeded with a small lambis, everything emerging perfectly, including the soft dark portion of its stomach. The operculum, or door, of a turbo shell is likely to crack in the deep freeze.

Little shells are difficult unless one can find a colony of friendly ants to do the work for you. Paraffin is also useful as long as it is changed frequently, so that the shells are not resting in sediment. When the contents are soft, syringe the shell with a hypodermic needle if ordinary tap pressure does not suffice.

Cones need wire implements and a great deal of patience to clean. I hook out as much of the interior as I can, and then turn them up with water covering what remains inside. For their periostracum, or outer covering, I soak cones in a caustic soda solution (one heaped tablespoon dissolved in a pint of water) for twenty-four hours. This loosens the covering making it easy to scrape the last murk away. Cones should also be dried and a little vaseline applied on the outside. If I intend keeping the periostracum by way of contrast, I grease it heavily, or it is liable to flake off.

I also oil my fragile bivalves, hammer oysters and date shells to prevent their outer skin from shrinking and shattering the shell. This can happen if the shell becomes too dry. Never add oil to a natural lustre; apply it only to a dull outside after having cleaned it with a knife or wire brush. Caustic soda is useful here for softening encrusting lime.

A decorated shell from the Seychelles.

Chapter 14

They Sailed the Eastern Seas

Although situated on the verge of the Indian Ocean, the Cape of Good Hope holds many of the bones of old ships which came to grief after crossing the Indian Ocean. With modern and sophisticated diving methods and equipment these wrecks are yielding their treasures and fortunes are being made by those who dare.

A Dutch Silver Rider brought up from the wreck of the *Meresteijn*. (Chris Jansen)

Sunken treasure: What magic these words hold. There are few people who at some time or another have not been captivated by legends of hidden Dutch East Indian silver, Inca gold or British bullion. Those of us who grew up on lashings of *Treasure Island* still hear, echoing faintly across the years, the screeching cry of Long John Silver's parrot: 'Pieces of Eight! Pieces of Eight!'

In a more modern period, those who have been prepared to carry through an undersea search to finality have often found their own hoards of pieces of eight; thousands of them that command a ready sale in the auction houses of Europe and America. More recent finds have included gold doubloons, silver ducatoons, delicate gold mohurs and star pagodas from India, George III guineas and sovereigns from a variety of British wrecks as well as a wealth of artifacts and jewels set in precious metal.

The ragged southern African shoreline has had more than its fair share of sunken treasure ships and rich legends of a bygone age. Over the past five centuries it has been estimated that more ships laden with valuable cargo have foundered in Cape waters than in any other comparably-sized stretch of coastline in the world. Long known as the Cape of Storms – stretching from Cape Columbine 100 miles north-west of Cape Town around Cape Point to Agulhas – the area has provided a watery grave to almost 500 ships; the majority either on their way back from the East or heavily laden with valuable trading cargo *en route* to some far-flung corner of yesterday's European empire. In an age of ultra-sophisticated electronic equipment and radar there are still ships which succumb to the weather or human error; how otherwise in a seaway which at the present time carries more than a million tons of fuel to Europe during each twenty-four-hour period. Today the gold is black. In previous years it was the real thing, and although part of this dangerous coastline lies across the South Atlantic all the ships concerned were headed for or returning from the legendary Indian Ocean.

The remains of one such ship uncovered in March, 1971 were those of a Dutch East Indiaman *Meresteijn* which went down off Jutten Island at the entrance to Saldanha Bay harbour on the evening of April 3, 1702. The *Meresteijn*, heavily loaded with cargo and passengers was on her way from Spanish Netherlands to Batavia. Her manifests show that she carried twenty-four chests of coins, mainly silver ducatoons, silver riders and stuiwers as well as a rajah's ransom in gold, all of which was to be used to finance Holland's garrisons in her Indian Ocean and Batavian possessions.

For almost three centuries the wrecked *Meresteijn* lay undiscovered and undisturbed until two intrepid diving brothers from Cape Town, Reg and Billie Dodds under the guidance of an ancient Saldanha islander Charles Adonis found her remains and brought up more than 1500 coins, as well as a number of finely-embossed bronze cannon and other artifacts; all within a period of a few months.

At the time of the catastrophe the Dutch were appalled at the loss. They tried reconstructing the wreck by sinking another vessel the *Wezel*, in the vicinity of the original mishap. All they managed to establish, however, was that the *Meresteijn* had initially proceeded over a sandy bottom. When the *Wezel* came to within 200 yards of where the other ship was reputed to have gone down an extremely rocky bottom was encountered. Heavy breakers made it impossible to proceed any closer inshore.

Under oath a report was submitted to the Dutch Governor of the Cape, Willem Adriaan van der Stel, setting out the findings and categorically stating that it would never be possible to recover anything from the *Meresteijn* as the huge seas and strong currents made any salvage venture prohibitive. Furthermore, owing to the steep gradient, it was said at the time, it was most unlikely that any of the specie would be washed up by the sea. An enormous fortune, conservatively estimated at today's rates in excess of £3 million sterling was therefore irretrievably lost.

Some twenty-five years later the undaunted Dutch East India

Company sent out to the Cape the famous English diver John Lethbridge who was requested to attempt a further salvage attempt. Lethbridge spent some time in the vicinity of the wreck, but later reported that he found conditions too hazardous to attempt any serious diving. He subsequently returned to England, but only after he had successfully retrieved a considerable amount of specie from a number of wrecks in Table Bay harbour – relics of other ships which had sunk in previous years in the shadow of Cape Town's famous Table Mountain.

Diving on the *Meresteijn*'s wreck is still no easy task today. Unless the sea is flat calm – which it seldom is, even in the best of weather – it is a hazardous and often dangerous undertaking. Only the most experienced divers have succeeded in getting anywhere near her.

Reg and Billie Dodds carefully examine booty brought up from the wreck of the *Meresteijn*. Note coins and cannon-balls in sack. (Chris Jansen)

A coin from the *Meresteijn* still coated with conglomerate while another is seen after cleaning. (*The Argus*)

Because the ship sank close inshore – the bulk of what remains lies between thirty and fifty yards from the precipitous shore – most excavations take place in the heavy surf which rolls in unimpeded across the South Atlantic. Consequently, the boats from which the divers operate must lie hundreds of yards to seaward; divers make their own way in, usually swimming underwater all the way to avoid being picked up by the rollers.

The entire hoard as it lies on the ocean floor is cased in a hard, black conglomerate which is chipped away from the surrounding bedrock with a hammer and chisel. This is placed in sacks and hauled on to boats by the tender crews. Once ashore the slow tedious process of sifting the booty begins. Individual coins are cleaned by an electrolytic process which removes most of the black stain, a feature of the *Meresteijn*, it has been figured, because she was carrying so much gunpowder when she went down.

One of the more interesting finds on the wreck was a ten-foot bronze cannon lifted off the ocean floor by two other local divers Kobus Steenkamp and Johnny Wilson. It took the two men two-and-a-half hours to raise this piece of ordnance, which weighed nearly 1 500 lbs. Smaller cannons taken from the *Meresteijn* fetched £1 500 among local art dealers.

Another Dutch East Indiaman lying nearby in the same bay is that of the *Middelburg*, which sank in shallow water on the morning of 1781. The *Middelburg* was one of five ships on their way back to Europe from the East heavily laden with cargo, including a number of crates of fine Ch'ien Lung china in her forward hold.

The Dodds brothers brought up one of these crates when they excavated the *Middelburg* during the late sixties. Together with coins and artifacts raised from the *Meresteijn* their finds were put

Chinese porcelain of the Ch'ien Lung Dynasty taken off the wreck of the *Middelburg* by the Dodds Brothers. (*Die Burger*, Cape Town)

on auction in Cape Town and Johannesburg in 1972. A total of more than £40 000 was bid for the various items. But it was not the first sale from the *Middelburg*, as in 1962 a complete Ch'ien Lung dinner set was sold by Sotheby's in London on behalf of the brothers for almost £30 000.

Hundreds of wrecks have been listed in southern African waters in recent years. Among the more interesting is that of the British brig *Fame* which was wrecked off Cape Town in May, 1822. The ship, under the command of Captain William Clark, left Madras in the Bay of Bengal bound for England by way of the Cape of Good Hope. After a voyage of over two-and-a-half months the ship arrived safely in Cape Town to take on fresh provisions and stores. She had hardly set out again when she foundered in a heavy north-west gale.

A week of pounding seas broke up the vessel until nothing more of the trim little ship was visible. For almost 150 years she too, like the *Meresteijn* lay undisturbed - broken and forgotten.

It was during October, 1965 after searching the sea bottom in the vicinity for several months that two other Cape Town diving brothers George and Jimmie Bell cam across the bones of an unidentified wreck. They were helped by the South African archives to identify her as the *Fame*.

Left is a British 'cartwheel' penny dating from the early 19th Century, here imbedded in conglomerate as taken off the *Fame*.

Gold jewellery also brought up from the wreck of the *Fame*. The centre piece is a brooch studded with diamonds and the larger item is part of a gold bracelet. The third piece is a gold dress ring. (Cliff Williams)

Using airlifts, pumps and waterjets, many tons of granite boulders, mud and sand were moved until some parts of the ship could be dislodged. These included a large gunmetal keel, the brass steering segment, a ship's bell and many copper nails. During these operations a number of small, black objects were seen and their weight suggested gold. When examined on the surface they proved to be 'mohurs', gold coins used as currency in India. Each bore the

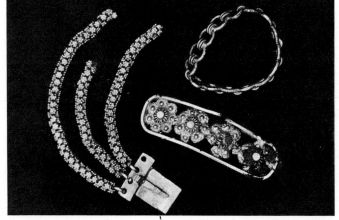

legend 'English East India Company' – under the auspices of King and Parliament.

A careful exploration of the site was then undertaken. Among the starfish, sea anemones and kelp lay scattered the remains of the treasure. Apart from star pagodas, mohurs and gold English guineas other finds consisted of pieces of eight and copper English pennies of that period; so many of them that they are still brought up by amateur divers working on the wreck today. Also recovered by the Bell brothers were silver spoons, handles of silver dishes, fragments of gold chains, gold finger rings and one ring with a central diamond and a topaz on each side. There was also a miscellaneous collection of unidentified brass and copper components and broken porcelain. Apart from the black deposit the gold coins were unaffected; the silver and copper coins showed evidence of corrosion.

Another interesting wreck lies only a few miles away from the remains of the *Fame*. This is the steamship *Hypathia* which sank off Robben Island when bound for Europe with a full cargo of copper and tin onboard. For decades these ingots have tempted divers; one man lost his life and I had a narrow shave while attempting to approach the vessel in heavy seas. The ship came to grief on a dangerous reef which is barely visible at low water and which causes tremendous swells and dangerous eddies to swirl around the wrecked hull. To add to problems the *Hypathia* is the haunt of seals, eels, the occasional shark and some of the largest octopi yet seen in Table Bay. One 'hard-hat' Navy diver who visited the wrecked steamer shortly after the last war had a twenty-minute encounter with an especially large octopus. He managed to cut off one of its tentacles which, when brought aloft, measured eight feet across. Stretched out the creature must have had a spread close on twenty feet.

Some of the bars of tin which a few of the more adventurous divers have managed to lift from the *Hypathia* were stamped 'Banca' the

Hard hat diver.

(South African Navy)

18. A cold-water Indian Ocean anemone. (Wolf Avni)
19. In the bowels of a wreck on the Moçambique coast near Lourenco Marques. (Johnny van der Walt)
20. A Grouper caught in the glare of a strobe light. (David Livingston)

18 ▲

19 ►

20 ▼

22
◄21
23
▼

name of an island in the Dutch East Indies. Much of the *Hypathia*'s original cargo remains although there have been a number of serious attempts at salvage.

Farther down the coast towards Cape Point, marked only by a tiny speck on a map of the Southern Cape, lies Albatross Rock which has probably claimed more wrecks than any other single position along the coast apart from the immediate vicinity of Cape Town. One of the ships which sank off Albatross Rock was the *Star of Africa*, a large steel-hulled ship with a considerable cargo of gold bullion and treasure onboard. One of the men who tried to broach her holds was the late Carl Ericson, a Norwegian who spent many years wreck-hunting off the South African coast. Ericson dived on dozens of wrecks but one of the worst he recalled in his later years was the *Star of Africa*. Employed by a Cape Town syndicate, Ericson dived on the ship during a brief calm period; like the entrance to Saldanha this area is exposed to the open sea. Ericson said that what worried him most about the wreck was that it swarmed with sharks. He counted sixteen large maneaters in the immediate vicinity of the wreck during his brief spell underwater. Although he tried to dynamite the wreck, the weather came up shortly afterwards and the venture was abandoned. The syndicate collapsed shortly thereafter.

Certainly the most intriguing wreck in any ocean is still the elusive *Grosvenor* which sank off the coast of Pondoland in 1782 after setting off from the port of Trincomalee with an estimated £4 million in treasure onboard. Her cargo included more than 700 gold and 1450 silver ingots, bags of gold coins, nineteen chests filled with diamonds, emeralds, sapphires, rubies and hundreds of pieces of ivory; a veritable treasure of the seas if there ever was one. This was not all she carried either. Her captain, a certain Mr. Coxon, wrote to his wife shortly before leaving Ceylon that he was bringing something 'that will set the whole of England by the ears!' Later reports indicated that the *Grosvenor* had onboard at the time of her sinking the Peacock Throne of the Grand Mogul that had

21. The diver goes cautiously along the propellor shaft of a wrecked freighter torpedoed off Lourenco Marques. (Johnny van der Walt)
22. Outriggers in the Nossi Be area. (Brian Rees)
23. Sunset off the coast of Madagascar. (Brian Rees)

once stood in the hall of the finest building in New Delhi, the Diwani-i-Am.

Part of the problem in salvaging this immense wealth of cargo is that no one knows exactly where the *Grosvenor* went down; she could be anywhere in an area covering several acres of rocky coast. The charts the ship had on board at the time of her stranding were old and out-of-date, while shortly before the calamity there was a difference of opinion between Mr. Coxon and his first mate as to exactly where they were. Coxon was of the opinion that the ship was 100 miles farther out to sea than they proved to be.

From the start an expedition was mounted in an attempt to find survivors from this wreck. Although some baggage and other items were recovered from an area where it was thought the ship went down nothing of value was found. Numerous skeletons were also found along the way. The rest of the survivors, it is believed, were either killed by hostile savages or assimilated by local tribes.

The first real expedition to recover the *Grosvenor*'s treasure was mounted by the Cape Governor seven years after the shipwreck. On this occasion only a few handfuls of coins, half-buried in the sand, were recovered. Two Scots, Alexander Lindsay and Captain Sydney Turner, had more luck some years later when they salvaged about 2000 gold coins while searching for the *Grosvenor*. This treasure was found some distance from the place where more recent expeditions have been active.

By the 1840s the British Government evolved a plan for recovering the *Grosvenor*'s treasure. A number of Malayan deep-sea divers were imported and although it was claimed that they found the wreck, it was covered with sand up to a depth of ten feet and nothing worthwhile was brought up.

The International Grosvenor Recovery Syndicate – the first of many such companies – was formed in 1905 to work the wreck, and although some coins were brought up, this venture crashed after their salvage vessel was dashed against the rocks and one of the divers killed. One version has it that the coins 'recovered' from the wreck were planted there in a bid to lure more financial backers for the venture. Shortly after the First World War the Webster Syndicate suffered similarly in spite of the fact that shares issued by the company sold like wildfire in Johannesburg when they were first issued. On this occasion a shaft was driven from the shore for

The author stands with two ammunition cases taken from the wreck of the torpedoed British ammunition ship *City of Hankow* which went down off the west coast.
(Chris Jansen)

Two Cape Town divers scour the bottom of False Bay looking for clues which will lead them to a wreck.

some hundreds of yards under the sea bed. Although some timber was found at a reasonable depth this venture too was abandoned soon afterwards, though no one could say why. Perhaps the money had run out.

The American millionaire Pitcairn entered the scene in 1921. More shafts were sunk and more money spent. Again the wreck yielded nothing. Then came a Dutch syndicate and thereafter a British group which seemed to have more luck than the rest for they brought up some gold and silver. But war intervened and the quest was again abandoned.

Since the last war there have been more attempts to recover the precious hoard onboard the *Grosvenor*; all have failed. But such is the lore of this fabulous ship that there are certain to be more attempts in the not-too-distant future. The question which must now be asked is whether any of them will ever succeed? Certainly the one that does will be rewarded beyond any man's imagination, for it is generally agreed that the *Grosvenor* was the richest of all the treasure ships that ever crossed the Indian Ocean.

A wreck somewhere north of Cape Town on a coast littered with the bones of yesterday's ships, most of them either on their way to or returning from the Indian Ocean. (Chris Jansen)

Chapter 15

Man's Deep Sea Friend: the Indian Ocean Dolphin

Gary Haselau, *former cameraman with the Jacques-Yves Cousteau television team onboard* Calypso *has spent many years studying or simply observing marine fauna in the Indian Ocean. Of all the creatures he has been associated with he regards the Indian Ocean dolphin as the most responsive and intelligent – and often an ally in a domain where man has few claims to acceptance.*

'In the reign of the late lamented Augustus, a dolphin that had been brought into the Lucrine Lake fell marvellously in love with a certain boy – a poor man's son who used to go from the Baiae district to school at Pozzuoli. . . . And when the boy called to it at whatever time of day, although it was concealed in hiding, it used to fly to him out of the depth, eat out of his hand and let him mount his back . . . and used to carry him when mounted right across the bay to Pozzuoli to school, bringing him back in similar manner, for several years, until the boy died of disease; and then it used to keep coming sorrowfully and like a mourner to the customary place, and itself also expired, quite undoubtedly from longing . . .'

Thus wrote Pliny the Elder of Rome a few years after the death of Christ, and ever since then man has been charmed by these graceful, friendly mammals of the sea who have shown a marked intelligence and amicability wherever they are found. So too it is in the Indian Ocean. There are many stories of friendly dolphins up and down the East African coast. They have also saved lives in the face of an imminent shark attack . . . but more of that later.

Few divers have not seen these sleek, swift models of gentleness at close quarters, usually playing in the bow wave of their boats. Others have had dolphins frisk among them while underwater, usually communicating with one another in short, sharp chirps or squeaks. Always the reception has been the same; one of curiosity at first, followed perhaps by a brief friendly romp before they move off again. Invariably they leave an impression of intelligence beyond the norm for certainly these are unusual creatures of the sea.

Dolphins are often referred to as porpoises, particularly in the United States. The name porpoise is apparently derived from *Porc-pes* or 'pig fish'; the cognomen referring to the blunt-headed

The bottle-nosed dolphin.
(Gary Haselau)

species of dolphin common in European waters during the Roman period. These porpoises are not usually found in the Indian Ocean waters but the name is in common use through this expanse of water. Because there is already a fish called the dolphin, as opposed to the mammalian dolphin, the name porpoise may avoid confusion.

Although there are more than fifty species of dolphins and porpoises, only six are commonly found in western Indian Ocean waters. Of these the common dolphin (*Delphinus delphis*) is the most widespread. During winter months they can be seen in large schools or pods off Cape Agulhas and Natal and in lesser numbers elsewhere. Also present in fairly large numbers is the bottle-nosed dolphin (*Tursiops truncatus*) which prefers the warmer seas east of Cape Town. They swim in pods of up to 100 strong and rarely move to warmer areas during winter months. It is this species of dolphin which is normally found in oceanariums as it readily adapts itself best to live in captivity. Both species inhabit the coastal or littoral zone of the Indian Ocean in preference to the open sea.

I enjoyed a unique opportunity to watch these creatures at close quarters from underwater during my voyage to America with Captain Jacques-Yves Cousteau on his research vessel *Calypso* a few years ago. I spent many hours in the submarine observation chamber at the bow of the ship watching bottle-nosed dolphins gambolling beyond the glass dome within feet of me and riding the pressure wave created by the ship's passage through the water. The

Calypso's pressure wave was only large enough to sustain one dolphin at a time so they made a game of it. When one dolphin had had its turn, another would zoom in and gently nudge the first one aside and take its place. While this was going on the others of the group would swim about with gusto, making clearly audible whistling bird-like noises.

Often the dolphin on the pressure wave would roll over on to its back, presenting its white underbelly to the surface. Looking down from above when this happened one could easily mistake it for an albino. Captain Cousteau told me that he thought this was probably how the legend of Moby Dick, the great white whale, originated. Whereas white whales are virtually unknown, whales of all species often sport on their backs showing their white undersides.

It was Captain Cousteau who first brought to my notice the fact that pods of whales or dolphins are invariably followed by one or more sharks. A pod of whales has been likened to a village on the move; the mammals are continually feeding, dying and giving birth leaving a trail of offal in their wake. Sharks consequently find it profitable to play tag with such a group.

One day this fact was graphically illustrated while *Calypso* was stalking a pod of pilot-whales with a view to filming them underwater. The whales knew they were being followed and suddenly doubled back, passing the ship a hundred yards away, and heading the way they had come. Accordingly we turned the ship to follow them. Almost immediately we spotted a large tiger shark coming towards us. It was swimming lazily at the surface on the same course the whales had been following. We disturbed the cruising shark and it took off at high speed away from *Calypso*. I have no doubt that it subsequently picked up the trail of the whales and resumed its endless patrol once again. The fact that we saw the shark – it was a large one – so close behind the whales and swimming in the same direction could be regarded as coincidental. However my voyage across the Atlantic Ocean and Caribbean Sea on board *Calypso* lasted four mid-summer months and in that time we saw only two sharks swimming on the surface of the sea. One was a small hammerhead which came to investigate our boat off Ascension Island; the other was the tiger we saw following the whales.

Apart from the fact that dolphins are airbreathing mammals, the main outward difference between these creatures and fishes is that, together with the other members of the whale family, the tail of the dolphin is horizontally orientated instead of vertical; this is of singular importance as it facilitates the dolphins' rise to the surface in order to breathe. All dolphins breathe through a blowhole situated at the top of their heads. Because they often travel at high speed in rough water they have evolved a quick-breathing system which can exhale and inhale between four and ten quarts of air in less than half a second.

After considerable research zoologists have concluded that dolphins were originally land animals, but for reasons unknown they returned to the sea about fifty million years ago. Their fore-legs evolved into fins for balance and steering; their rear legs have all but disappeared with only vestigial bones remaining enclosed in the body. The dolphin's tail developed into powerful flukes to propel itself through the water; these flukes consist only of cartilage and contain no bone structure at all.

So far as food is concerned an individual dolphin requires about 15–20 lbs of fish each day. Because of their intelligence and sonar capabilities a pod of dolphins has no difficulty in locating shoals of fishes, surrounding them at will and picking off as many as are needed by the group. It seems likely that the dolphin gets its necessary supply of fresh water from the flesh of the fish it eats. Their jaws open very wide and contain eighty-eight sharp, pointed teeth.

Dolphins have excellent eyesight both in and out of the water. How they accomplish this remains something of a mystery. Apparently the shape of the lens of the eye can be altered at will to correct for either type of vision. They can retrieve an object high above the water's surface which they home into from underwater; they invariably catch objects thrown to them while the upper part of their body is out of the water. Their eyes are situated at the widest part of the head enabling them to see in all directions and also giving them binocular vision in the forward direction.

The Indian Ocean dolphin emits a great variety of sounds. To the human ear these sound like squeaks, grunts and whistles. It is certain that they are able to communicate vocally with each other and it has been shown that they use certain of their sound emissions as a form of sonar or echo-location for navigating purposes. Tests have proved that their sonar abilities achieve a high degree of accuracy in opaque or murky water conditions; to do this the dolphin 'sounds' ahead in a series of high-speed clicking noises. The pitch of these clicks varies from the lowest audible sound up to at least

Bottle-nosed dolphins at play with a diver off the South African coast. (Gary Haselou)

170 000 cycles per second (this is about ten times beyond the range that man can hear unaided). The echoes of these clicks returning to the dolphin indicate by time-lapse and other variations where it is in relation to its surroundings.

To communicate with its mates the dolphin emits whistles in about the same pitch range that we can hear. An amazing fact is that the dolphin can both talk to its companions and use its echo-location clicks simultaneously.

The accuracy of this echo-location system can best be illustrated by an experiment carried out in Hawaii. It was found that a blind-folded dolphin could tell the difference between two balls; one two-and-a-half inches in diameter and the other two-and-a-quarter inches in diameter. At the time the experiment took place the dolphin was about six feet away from the balls.

Physically the dolphin is powerfully built. One scientist has estimated that its muscles are seven times stronger than those of man. This, together with its superbly streamlined body, enables it to travel at speeds of at least twenty knots in the water, and jump twenty-five feet into the air. Water resistance is reduced too by a rubbery skin undulating and forming flow-curves which considerably reduce friction as it moves through the water.

At rest the dolphin can stay underwater for about six minutes, although when moving at high speed it takes a breath every twenty-five to thirty seconds. We do not know for sure how deep a dolphin can dive but I estimate an ability of at least 600 feet.

Dolphins normally mate in the springtime; the gestation period is between eleven and twelve months. At birth the baby bottle-nosed dolphin weighs between 30–40 lbs and is approximately three feet long, compared with the adult bottle-nose which weighs approximately 400 lbs and is six to nine feet long.

What is unusual about the actual birth is that the baby is born tail first. Immediately after birth the baby has to be lifted to the surface in order to breathe. If it was born head first in the normal manner of all other mammals the baby would drown before the birth was complete.

When suckling her young the female has the ability to squirt her rich milk under pressure into the baby's mouth so that the nutrient does not mix with surrounding sea water. The baby then surfaces to breathe before receiving its next mouthful. Young dolphins stay with their mothers for up to three years after birth.

Scientists are intensively studying the complex brain structure of the dolphin which is large in comparison to its body size. According to studies already undertaken the dolphin's brain is equally if not more complex than that of man and even larger. While the human brain is approximately 1 400 grammes in weight that of the dolphin averages 1 700 grammes. Obviously there is more to this playful creature than first impressions decree; because of the animals dexterity and intelligence it will be interesting to observe the results of future experiments. Some scientists believe that some form of communication between man and dolphin is possible; in fact one Port Elizabeth scientist has claimed that he has already achieved some success in this direction. (Late 1972.)

One of the most famous scientists working in this field is an American, Dr. John C. Lilly, who has set up two laboratories for the study of dolphins and their methods of communication; one is on the island of St. Thomas in the Virgin Islands, while the other is at Miami in Florida.

Before becoming interested in dolphins Dr. Lilly practised for some years as a neurophysiologist, studying the brains of cats and monkeys in relation to man. He was aware that in a developing human child the brain needs to weigh at least 1 000 grammes before it is capable of speech. Because of the size of the dolphin's brain in relation to its body, Dr. Lilly concluded that it was perhaps possible that dolphins were able to develop an advanced form of communica-

Tersia Venter, wife of the author, with in the background a converted Arab dhow, which is one of the strange and wonderful diving launches the Venters used in the Indian Ocean. (Al Venter)

tion among themselves; and possibly with other intelligent beings. Further study showed without doubt that dolphins do have a language of some kind. Accordingly, he felt, if there was any chance at all of man being able to communicate on an intelligent level with an alien species the dolphin would be the most likely candidate. Since then Dr. Lilly has made significant progress with his research, and has learnt a great deal; unfortunately, space limitations preclude full comment here, but I will refer to one incident which gives insight to the dolphin's inherent capabilities.

One of Dr. Lilly's first projects was to record as many different dolphin sounds as he could, as a basis from which to commence his work. He isolated one of his dolphins in a smallish tank and began working in the normal way of rewarding the animal when it did something right and ignoring it when it did not respond correctly.

Soon the dolphin realized that Lilly was interested in the sounds it could make. Of its own accord it obliged by emitting all manner of noises, and each time it did something new, it was rewarded. Lilly noticed that the sounds he was getting were becoming progressively higher in frequency until he could no longer hear the dolphin, even though he could see by its behaviour that it was still vocalizing. At this point he stopped providing further incentives. Immediately the dolphin sounds came down in frequency to where he could hear them once more; again Lilly provided adequate reward. The dolphin had then established the range of Lilly's hearing and did not exceed the range again during that particular experiment.

The United States Navy has carried out a great deal of research into useful work that dolphins and other toothed whales could possibly carry out on behalf of man. During one of their sea-lab projects, where teams of divers lived in an undersea habitat on the ocean floor for weeks at a stretch, a bottle-nosed dolphin named Tuffy was trained to take messages, mail and tools down to the men living 205 feet below the surface of the sea. Tuffy also acted as a safety guide for the divers. Because the divers were living under pressure-saturated conditions they could only return to the surface at the end of their stay underwater after long periods of decompression. If a diver became disorientated on the sea floor and was unable to find his way back to the habitat, he would then have been faced with serious problems. His air supply would have soon been exhausted, and because of pressure limitations he could not surface as he would certainly die painfully of decompression sickness (the bends). Tuffy was taught that if a diver became lost he would sound a buzzer which he carried with him. On hearing this particular sound the dolphin would know what was happening and would dive straight down to sea-lab. There Tuffy would pick up the end of a life line on a reel attached to the habitat and using his sonar to find the diver in the dark or dirty water, would take him the life line. The diver would then simply follow the line back to the habitat.

The U.S. Navy has conducted numerous other experiments with dolphins, some of which are top secret. They have trained a larger relative of the dolphin, the pilot whale, *Globicephala melaena*, to recover practice torpedoes which are often lost on the sea bed. These recoveries have been carried out in depths of up to 1 200 feet. The whale simply attaches a recovery device to the torpedo which is then lifted to the mother-ship above.

Having had a wonderful relationship for many years with a bottle-nosed dolphin called Haig at the Port Elizabeth Oceanarium, I can vouch for their friendliness towards man. There are numerous anecdotes regarding this spontaneous and uncanvassed amicability towards human beings.

Two of the best-known relationships of recent times have been that of Pelorus Jack and Opo, both New Zealand dolphins of world fame. Pelorus Jack was a dolphin which accompanied ships across Admiralty Bay between Wellington and Nelson. Tourists often took this cruise just to see Pelorus Jack, and he became so famous that a special law was passed in the New Zealand Parliament to protect him. He disappeared in 1912 after three decades of continuous performance.

Opo was a female dolphin which suddenly appeared one day in 1955 off the beach of the small town of Oponoric. She preferred the company of children and often played ball with them. Sometimes she would let some of the smaller youngsters ride on her back. As Opo's fame spread abroad people travelled far to view her antics, and once again a special law was passed to protect her. When she died she was buried near the beach and a monument erected over her grave.

About twenty years ago two bottle-nosed dolphins appeared off the beach at Fish Hoek near Cape Town. They made a habit of playing with the bathers and were promptly named Fish and Hoek. At that time I had just begun to take underwater pictures and I attempted to take shots of these two rompers. However, as I was still very much of a novice in underwater techniques the photographs were not much good. I did, however, hitch a few rides by hanging on to their dorsal fins. Fish and Hoek would tow us about twenty yards before tiring of the game and shrugging us off. Then one day they departed and were never seen again.

There are many tales too of dolphins saving the lives of people cast adrift at sea. Unquestionably these actions have been positive and well-meant. Not long ago a South African fisherman claimed to have been saved from drowning when his boat sank near Cape Town. He said that a dolphin had supported him for hours until he was rescued. More recently, in August, 1972, a pleasure boat over-turned in heavy weather near Lourenco Marques in Moçambique. A young girl by the name of Yvonne Vladislavich found herself in the water twenty miles from shore, and there were sharks about. Yvonne recalled later that her feet were bleeding, probably from injuries received when the boat sank.

At one stage Yvonne found herself in the company of a number of sharks. At the crucial moment – as the sharks were closing in – two dolphins appeared from nowhere and kept the predators at bay. Dolphins despise sharks and often kill them by butting them at high speed in their vitals.

For a long time the mammals stayed with the swimmer. Whenever Yvonne showed signs of weakening the dolphins would nudge and apparently support her until she had regained her breath again. They stayed with her until she reached a large shipping buoy from where she was rescued early the next day.

There are numerous similar cases recorded in the annals of the sea and although some folk choose to believe that many are based on fantasy, I personally feel that in these cases dolphins understand that here is a fellow creature in distress, and they therefore support the castaway, just as they do with their own kind.

I also prefer to believe that having the intelligence that they do, they instinctively realize that we are land creatures. This would explain why they often, of their own volition, push survivors towards the shore.

If you should doubt this, conjure up an image of a highly intelligent creature with a brain more complex than your own and larger; a brain capable of intelligence beyond our present comprehension . . .

Chapter 16

East Africa's Marine Parks

Kenya, in East Africa, has taken a lead in protecting her maritime heritage from tourist plunder; and none too soon. Already this coast has attracted some of the world's best-known undersea enthusiasts, including the Cousteau family. The underwater attractions are multiple and fascinating, from the legendary Arab town of Lamu in the north all the way down the coast to Tanzania and the Moçambique frontier.

Sign on the main road between Malindi and Watamu.

On a continent fraught with multiple political and economic imbroglios it comes as something of a surprise to learn that Kenya took the lead in establishing Africa's first series of Marine Parks. The notice in Kenya's *Official Gazette* in March, 1968 declared that 'an area of the Indian Ocean within Kenya's territorial waters adjoining the south-eastern boundary of Malindi Town . . .' and another area at Watamu a few miles to the south shown on a map and 'delineated in purple' were to be declared a Marine National Park with immediate effect.

Other areas which have since obtained quasi-protected status under the country's conservation laws include a holiday zone near Kilifi on the main road south as well as some coral island reefs off

The Indian Ocean area has thousands of quiet inlets and lagoons which make for ideal diving conditions. This one is situated south of Kilifi in Kenya. (Al Venter)

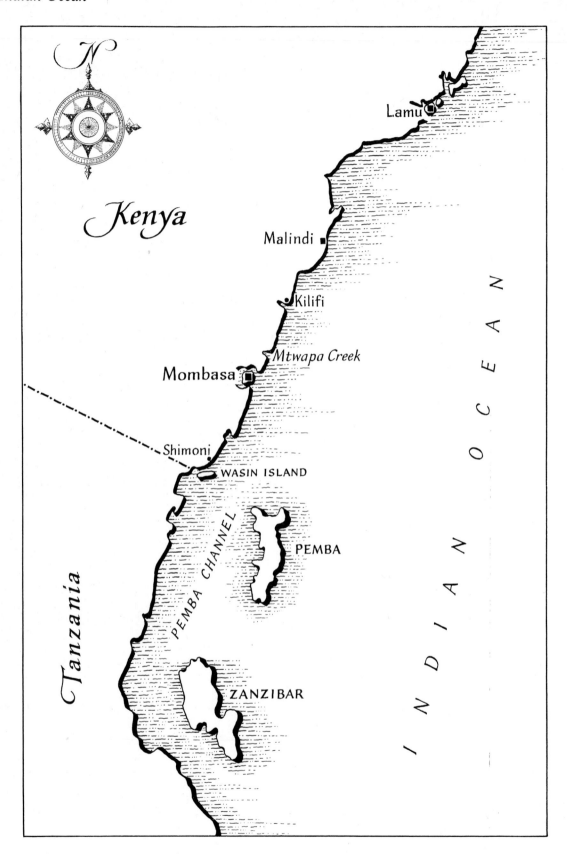

Shimoni towards the Tanzanian frontier. Local residents who know Shimoni well are thankful that this area – although remote from the usual tourist haunts – has been taken under the mantle of Kenya's Ministry of Tourism and Wildlife, for the range of coral and marine life here is regarded as of the best to be found anywhere off the coast of Africa.

On her visit to Kenya in 1971, Britain's Princess Anne was taken on a brief skindiving sojourn in the Shimoni area. Experienced television crews who accompanied the Princess regarded the area as the finest they had ever seen.

The Marine Park proclamation came none too soon. It provided 'instant safety' for a large variety of marine life which were becoming increasingly popular in the international tropical fish market. A number of Kenya residents had started their own tropical fish exporting businesses and within a year the depletion could be readily noticed, especially in the Malindi area.

The Kenya coast is well supplied with diving facilities such as this equipment which is available at Malindi. (Al Venter)

There was also considerable demand for East African cones and cowrie shells which were being sold through local and overseas outlets in such quantities that by 1966 it was observed that the region's ecology was being disturbed. All sea-shells, in their own peculiar way, serve a useful purpose in maintaining the balance in any ocean area. So many large conch and triton shells had been removed from Malindi waters by 1968, for instance, that the crown-of-thorns, *Acanthaster Planci*, menace which had until then been kept in check by these huge shells which find a ready market among tourists, became as serious a problem along parts of the East African coast as it already was along Australia's Great Barrier Reef.

The system of control within Kenya's Marine Parks is simple. Covering an area of approximately seven square miles the region is patrolled by a number of uniformed wardens who operate on foot along the beaches and from boats based at Malindi and Turtle Bay. Two offices issue tickets to anyone wishing to enter the area, slightly less being paid by Kenya residents than by tourists (a standard procedure in all Kenya's Game Parks). The chief warden is based at Malindi and by 1972 his staff had grown to thirty. Two powerful boats were provided by the Nairobi Government to help in his daily task.

Restrictions were successful from the start. Although poachers – mostly unauthorized African fishermen – were arrested from time to time, marine life in all controlled areas has increased substantially in the interim period. Controls have also had a marked effect on the normal habits of fish encountered in the Turtle Bay area. Visitors who are taken out by tourist launches are surprised to find how friendly and how numerous sea life is, ranging from hundreds of jewel-like *koli-koli*, *Caranx* family, and spotted *futi* (gaterin) to the

Spotted gaterin (*Gaterin gaterinus*) off Watamu beach. (Peter Saw)

occasional 300-lb. humped-back *tewa* or grouper which inhabit the Big Three Caves at the entrance to Mida Creek.

Enter the water and the skindiver is immediately surrounded by fish which will follow groups of swimmers around between the coral heads for hours at a time. The same applies to some of the larger species, including a good few large green turtles and smaller hawksbill turtles. In 1970, a female turtle in the Watamu area became so friendly with a local resident, Mrs. Pauline Saw, that she regularly allowed herself to be played with each time Mrs. Saw entered the water. Eventually the reptile would recognize Mrs. Saw and approach her of her own accord. After a few weeks of association Mrs. Saw found that her hard-domed friend had disappeared; she concluded that the turtle had probably laid her eggs and moved on.

Pauline Saw plays with a friendly turtle in the Malindi Marine National Park. (Peter Saw)

Possibly one of the most interesting areas within the Kenya Marine Park complex is Mida Creek, a huge expanse of partially closed

Exploring an East African
undersea reef.

Known in Kenya waters as an 'old lady', this
tropical fish is found throughout much
of the Indian Ocean. (Peter Saw)

Indian Ocean soft coral.

Spotted rock cod in a coral overhang. (Peter Saw)

One of the best reasons for declaring an area a marine park can be seen from this photo taken by Peter Saw of his wife Pauline in the Malindi area.

Indian puffer fish. (Peter Saw)

Anemone and a clown fish: each dependent upon the other. (Peter Saw)

lagoon which empties out with the tide through a narrow channel linking it to the sea twice each day. Several square miles in area, Mida Creek consists mainly of coral garden and acres of sea grass of endless variety and contrast. Visitors are advised to enter the water at the inshore end as the tide recedes and allow the water-flow to carry them across the coral heads towards the entrance of the Creek. As the water moves at about two knots this allows ample time to explore numerous crevices and gullies in the vicinity although the uninitiated should keep their eyes scanned for the occasional sea snake which, it is said locally, breed in the upper creek shallows.

But it is at the entrance to Mida that a bonanza awaits those who have stayed the distance. Here, in an area several hundred yards out from shore in the middle of channel, lies one of the rare experiences in any diver's life – the cathedral-like Big Three Caves which can only be reached during that brief half-hour when the tide is on the turn. The flow of water through the channel is too strong to allow for diving within the tunnels and openings during the normal movement of the tide.

Within a few feet of the surface a number of large openings are to be found on the top of the reef. Appearing rather ominous at first glance, they lead downwards perhaps ten feet and then open up into huge subterranean chasms which contain a fascinating variety of marine life including some of the largest giant sea bass or *tewas* along the coast; the Indo-Pacific giant sea bass or grouper can grow to twelve feet in length and weigh as much as a ton.

At the last count the grouper family – in the shadows the largest of these fish appear almost as vast as a small Italian car – numbered about eight; five more than there were when the park was first declared.

The *tewa* family and its attendants – there are moorish idols and brilliantly hued surgeon fish by the hundreds – as well as a good few morays which turn the most experienced diver in their tracks – have never been known to present any serious problems to those who explore their domain. The *tewas* are naturally curious and approach most divers who enter their cavern with more than casual interest. Unlike the West Australian species of grouper the East African variety have never been known to attack divers; it could be that East African *tewas* are better fed. Although the caves lie on a reef some distance from the shore they are not within the lagoon and therefore some caution should be observed for sharks. There

Dried seacat (octopus) is a
great favourite among Africans
in the Indian Ocean area.
This was on Quifuti Island.
(Margaret Smith)

have been no attacks to date, but lone sharks have been caught on
the nearby seaward side of the reef.

What is surprising considering the success of the Kenya Marine
Parks venture is that there was considerable opposition to its
original establishment. From the start it was accepted that a number
of local African fisherfolk were dependent for their livelihood on
what they caught inside the reef area, but on declaring the area a
protected zone some concessions were made to these people, who
are usually discriminative in what they catch.

What has been eradicated is large-scale and indiscriminate spearfishing which became endemic with the arrival of the first German
tourists in the area. Before the establishment of the reserve a
number of German tourists, complete with scuba gear and spearguns, had set out to eradicate the Big Three Cave *tewa* community.
They were only stopped by one of the white local residents who
heard of their venture and showed his disapproval by firing over
their heads from the shore with his hunting rifle. The *tewa* hunt
was stopped but not before one of the largest specimens had been
shot and left to die in the caves.

In the past it was not unusual to see the butchered remains of
female turtles and evidence of their demolished nests. As with the
Zululand coast farther south, the Malindi area is also a regular
breeding area for a few of the Indian Ocean varieties of turtle.
Although this activity has not been completely eradicated it is
controlled at present and some clutches of turtle hatchlings have
reached the sea in recent years.

There has also been considerable interest in Kenya's Marine Parks
in Tanzania, south of the border. Following the establishment of
the Kenya venture, the Tanzanian Government invited the noted
American zoologist, Dr. Carleton Ray of the Johns Hopkins
University, Baltimore, to investigate the offshore waters of this
corner of the Indian Ocean and determine their 'suitability for a
national park'. With considerable experience in this field behind
him – it was following his pioneer work that the Exuma Cays
Land and Sea Park was established in the Bahamas – Dr. Ray was
allowed *carte blanche* throughout the length of Tanzania's 1 000-
mile coastline, which stretches as far south as Moçambique.

Dr. Ray's findings are important for they are applicable to the
entire Indian Ocean littoral zone.

East Africa's Marine Parks

Two wardens of the Kenya Marine Parks at Malindi. (Al Venter)

A girl on the ocean floor. (Conway Plough)

African fish nets are covered at high tide, when some interesting catches are made in these primitive though functional devices. (Al Venter)

Examining at close quarters underwater life. (Peter Saw)

A Kenya Marine Parks patrol boat does the rounds in Watamu lagoon. (Al Venter)

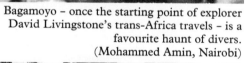

Bagamoyo – once the starting point of explorer David Livingstone's trans-Africa travels – is a favourite haunt of divers. (Mohammed Amin, Nairobi)

Working off various tropical shallow and deep-water reefs in the territory Dr. Ray found that in spite of a relatively undeveloped tourist market Tanzanian reefs were being rapidly destroyed by shell and coral collecting, dynamiting by fishermen, over-fishing and trampling by net fishermen, silting caused by upland deforestation and other factors. The lack of a modernized fishery industry, he found, was in part responsible for these practices which were destroying habitat for fishes and other marine organisms, thus reducing productivity.

In spite of these negative developments, Dr. Ray reported that Tanzania's marine resource was rich in content and extent. Some of its coral reefs, he said, were among the finest on Africa's eastern coast and compare extremely well with those in other tropical areas. Some, he maintained were 'spectacularly unique'.

There were a few more disturbing pointers. Some species such as green turtle and dugong – a large marine mammal which is related to the manatee – although legally protected, were being excessively hunted. It was Ray's opinion that they were in danger of extermination. (See photograph of dugong harpoon.)

A spear used to harpoon dugongs off the East African coast.
(Peter Saw)

Shellfish, he also concluded, were becoming rare due to excessive collection for the shell and tourist trade. This activity threatened to deplete the very resources which the tourists came to Tanzania to find.

Sadly, although his recommendations were made more than five years ago, no positive action has been taken, mainly, it has been reported by some observers due to obscurantism within the responsible government department. Dr. Ray suggested that a number of marine parks be established over extensive areas along the Tanzanian coast. He felt that some of these should serve as 'buffer areas' around the projected preserves.

One of the comments made by Dr. Ray concerned the common Indian Ocean practice of dynamiting coral reefs for the purpose of stunning fish, which causes them to float to the surface where they are collected by local fishermen. The measure, though highly illegal, is a shortcut to a fairly substantial catch if there are fish shoals in the area.

But, Dr. Ray warned, at the present rate of destruction he estimated that it would take less than a decade to result in the complete

Dar es Salaam harbour, once considered the best spearfishing area along the East African coast. (Al Venter)

expiration of all coral reefs in the Dar es Salaam area. Because reefs were important to fish as cover, the use of explosives had the effect not only of reducing that cover but also killing all fish within a radius of the blast. The overall result was a reduction in tourist appeal. Further, when a reef was flattened by an explosion, the elevation was also destroyed; silting covers the bottom where new attachments of coraline organisms would otherwise occur and many years are thus necessary for recuperation.

A species which Dr. Carleton Ray sees in serious jeopardy is the *tewa* or giant sea bass, and here he holds indiscriminate spearfishermen to blame. He says: 'These huge fishes live in shallow water accessible to the diver (off the East African coast). They are much sought after by the spearfishermen who kill them for sport, but can rarely utilize their 200–2 000 lbs. of meat.' He also warned that these fishes can be dangerous.

Six areas, in particular in the offshore Tanzania area, were recommended as future park and reserve zones. These included the Tanga coral gardens which are really an extension of the Shimoni reef; the Maziwi Island area, regarded by Ray as a 'spectacularly beautiful island' which is also surrounded by fine coral gardens and

Feeding fish by hand in a
Kenya Marine Park.
(Peter Saw)

Among the bamboo fish nets
at high tide. (Peter Saw)

Arab coral anchors, such as this one
which is four feet high, are regular finds
in many East African ports. (Al Venter)

is used as a breeding area by green turtles, *Chelonia mydas*, besides the Dar es Salaam area, which includes the islands of Mbudya, north of the city, and Sinda to the south.

Latham Island, an uninhabited bird sanctuary forty miles east of Dar es Salaam, was cited as a fine example of a joint bird colony and marine sanctuary, but it was suitable as a park only if the necessary measures were taken. Although Latham is not a true oceanic island it represents Tanzania's nearest approximation to an atoll by virtue of its isolation. its surrounding deep, its blue waters and its oceanic bird populations. Mafia Island to the south was also recommended as a controlled area, although tourism in this vicinity has developed substantially since the report was published.

Certainly the most interesting recommendation was the suggestion that the Kilwa area – an ancient Arab settlement dating from the 10th Century – be declared a dugong reserve.

Kilwa is known to be one of the last remaining areas where the mermaid *dugong dugong* is still relatively abundant. Though the species is protected by law, hunting of this mammal continues.

The dugong is a peculiar beast. Together with the manatee both gave rise to ancient tales of mermaids due in part to the fact that the female has two breasts on the chest rather than on the pelvic area, as in all other animals except elephants and primates. The Tanzanian and Moçambican dugongs are vegetarians which inhabit murky coastal waters. The flesh is excellent and, Ray maintains, there is a distinct possibility that it could be farmed though the reproductive rate is low. A rare animal, many of the dugongs habits remain to be studied. For a diver to spot one is still a unique experience; to photograph one highly unusual.

Two Kenya divers attached to a 'hookah' harness dive off a reef north of Kilifi.
(Peter Saw)

Chapter 17

Mauritius Undersea Venture

Known to the mariners of old it is only recently that Mauritius has become part of the jet-set circuit. Royal Navy Lieutenant Commander **Bill Abbott** *tells of some of the diving lore off this volcanic core in the middle of the Indian Ocean.*

The venue is Mauritius: Pearl of the Indian Ocean and a dozen other equally flowery epithets. The image is fine but the symbol had become just a little jaded as the tourist flow to this erstwhile French and British island-colony increases.

But it's not just the visitors who have tarnished what was once a fine tradition, for it was locals – amateur diving enthusiasts and professionals alike – who started dynamiting the island's abundant coral reefs and continue to do so in spite of the threat of severe penalties if caught. It was also local residents who have played a prominent if rather ignominious role in cutting down the variety of marine life to be found in offshore waters through indiscriminate spearfishing. Here too the authorities had to step in smartly. For this reason only foreign tourists may use a speargun in Mauritius waters.

But problems aside – and there is not an area in the world without them in a global society stretching ever outwards in its determined search for the exotic – Mauritius remains a minor paradise in its own right. This much is clear even as the aircraft makes its initial approach run over Le Morne on its way down to Pleasance Airport. Clear water and tropical reefs glimmer upwards in a shower of turquoise and blue. Light shades grow darker on the verge of the oceanic shelf where the sea suddenly drops away to eternity and beyond.

But to dive in Mauritius you have to know where. For those un-initiated in the vagaries of an Indian Ocean monsoon season or the full force of a Mauritius south-east 'blow', a diving holiday can end in mishap if the enthusiast is not correctly advised.

For a start, most locals prefer the west coast, around Albion, Flic-en-Flac and Black River areas where water visibility on a fair day rarely falls below 100 feet.

Unlike Madagascar and the coast along East Africa, most Mauritius 'seafaris' take place *outside* the coastal ring of reefs from boats which are easily hired locally with the assistance of local divers or one of the island clubs.

One of the most popular venues is at Trou Aux Biches, where the depth is about sixty feet, there is good visibility and little current. Boats usually anchor about 200 yards beyond the reef in sheltered waters when the weather is fine. Only a slight ground swell provides a reminder that the diver is in the sea and not in an enclosed stretch of water.

A feature of diving off Mauritius is the coral; enough brilliant forms and colours spread out among intervening stretches of white sand to make for a backdrop of any undersea wide-screen scenario. Because of the variety, a first dive at Trou Aux Biches is as good an introduction as any, although some form of body protection – either a wet-suit or a shirt – should always be worn because of coral stings.

For those who know where to look Mauritius also has its own rare black coral. A few isolated colonies of black coral have been discovered in deep water off the north-west coast but the actual location is kept secret by local residents who fear that the speculators may not be conservation-minded enough to realize that once depleted the delicate black coral may take generations to grow again to maturity.

More accessible than Trou Aux Biches is the Ambulante Passe which lies just off the Le Morne Hotel. The Passe is one of the deeper natural channels in the encircling reef through which the marlin boats daily set out to sea in search of larger game-fish – marlin, tunny or shark. This area is excellent for crayfish-diving before breakfast and some of the larger specimens of crustaceans brought up have tipped the scale at eight pounds.

The Passe promises excellent spearfishing for those visitors who wish to try their hands. Maximum depth in the area is about forty feet, with water of varying clarity ranging between twenty feet and 100 feet depending largely on the state of the tide.

Delightfully coloured moorish
idols on a reef hardly noticed
a visiting diver. (Peter Saw)

Fish consist mainly of streaker, a species of kingfish, parrot fish
and dog-toothed tunny. Interestingly enough, local professional
divers who know the area well say that sharks are very rarely seen
in the area although a Durban diver, Eric Roest, came face-to-face
with what he termed 'an obviously hungry' Java shark on his very
first morning in the water. Each morning and evening that Roest
spearfished in the area the shark maintained its silent, distant-
circling vigil but never attacked. According to Roest the shark was
quite large and could easily have caused problems had it wished to
do so.

The islets to the north-west also afford great sport, with abundant
supply of wahoo, tunny, marlin, caranx and barracuda in water of
crystal clarity. Diving in this area, however, is not for the novice;

Coral gardens off Trou
Aux Biches, Mauritius.
(Peter le Lievre)

the islands lie exposed to the ocean swell and tidal streams can cause the occasional dangerous race.

A deeper and more spectacular dive than that at Trou Aux Biches or the Passe is to be found near a village called Flic-en-Flac. At a depth of sixty feet the black basalt reef ends abruptly and forms a potholed cliff face with surprisingly little coral on its surface, descending vertically to 100 feet. At the base of this sheer drop divers have discovered a large cave with an entrance about ten feet across which they have significantly named 'The Cathedral'. The description of the inside of the cave is apt; the interior is about fifty feet long, with an arched roof ten feet high. Natural light inside the 'cathedral' is sufficient to see the unblinking eyes of a large 'congregation' of rock cod, invariably fixed on the intruding divers. It is possible to sit on the floor of the caverns and watch silver bubbles of air collect under the roof, eventually to percolate away through tiny holes, ultimately emerging from the reef at sixty feet. At the farthest end of the cave there is a shaft, or 'chimney', about six feet across, which ascends vertically about sixty feet to the reef. Ledges in this shaft have become the lairs of numerous large moray eels which should be treated with healthy respect by anyone attempting the ascent. They have in the past been known to leave their holes to attack if provoked and among them are a few lethal 'critters' well in excess of eight feet long.

Cliff faces in Mauritius are regular haunts for the occasional solitary cruising shark, and visitors must be prepared for contact – if only fleeting – on a dive in this area. Mauritius sharks are generally well-fed and benign but they have a tendency to become curious and often circle divers for various periods. They have never been known to attack since diving first became popular here in the sixties.

Sharks will, however, take a spearfisherman's catch if the opportunity should present itself. All speared fish should, therefore, be attached to a length of line on a float. Shoals of barracuda are also commonly seen in this part of the Indian Ocean and they can present problems to the novice by their fast, erratic movements at close quarters. Like sharks, barracuda tend to circle divers, and there is little one can do, except form a defensive circle and ascend slowly to the boat. In each case rapid movements, or a counter-attack by speargun are not recommended.

Both these predators highlight a problem which can have serious consequences if depth limitations are not assessed and enforced from the start by anyone diving in Mauritius waters.

Most diving spots off the island are to be found in deep water and because of the unusually clear water and tropical undersea panorama one tends to lose sight of depth and time; this can be an

Coral, beautiful fish and very clear sea are all qualities offered by Mauritius.

extremely hazardous failing in an area where the nearest decompression chamber is more than 1 000 miles away. Visitors to Mauritius are advised to keep a careful check on their own capabilities and limitations, for a case of the 'bends' can have fatal consequences. Three experienced divers died in Mauritian waters in this way during 1972.

However, diving in Mauritius is not all sheer cliff faces and sharks. The island boasts some of the finest accessible wrecks to be found in the Indian Ocean and no visit would be complete without a visit to some of them.

One of the more interesting hulks lying within easy reach of the shore is off Mahebourg on the south-east corner of Mauritius. This is the *Sirius*, sunk in the action of the Battle of Grand Port in August 1810, and to this day her cannon can be seen lying in neat rows on the bottom at about 100 feet; these are surrounded by numerous irregular hunks of timber and copper sheathing, much of which has been removed by visiting souvenir-hunters and scrap-merchants. Many other relics of this famous naval battle are to be seen in the Mahebourg Naval Museum.

Of literary fame, the well-known French novel *Paul et Virginie*, by Bernadin St. Pierre, is set in Mauritius and concerns the wreck of the *St. Geran* which lies in twenty feet of water off the north-east coast. Some pieces of eight can still be found by those willing to look among the shapeless lumps of coral and conglomerate, while her anchor and cannon are all visible. A dive on the *St. Geran* involves a long boat trip and must be undertaken in fine weather; the wreck lies off an exposed coast and it takes the best part of a day to complete the expedition, but the journey is well worth the effort.

Mauritius resulted millions of years ago from a mighty volcanic eruption on the ocean floor and like nearby francophone Reunion it lies off the main sea routes to the Persian Gulf. They are therefore not subjected to pollution like parts of the East African coast.

Erosion caused by a fairly high rainfall and the crumbling nature of the volcanic rock has produced a fairly broad shelf of debris around the island upon which numerous coral reefs have developed over the years. Compounded by various factors a variety of shallow lagoons have formed behind the shelf providing even the most elementary diver with a fascinating display of marine life. Here, in crystal-clear water the snorkel diver will find any variety of angel fish, parrot fish, quaint and colourful *demoiselles*, target fish and the largest variety of eels in any Indian Ocean coral zone. These include smaller moray eels and the harmless but often alarming banded snake-eel which often swim directly for any diver who has disturbed its lair.

The island is also noted for its shells; conches, cowries, cones, some common and a few rare. All can be found with a modicum of patience, in fairly shallow water.

A favourite spot for the not-so-enthusiastic is found near Black River in the south-west. Stretching from Morne Brabant to the Tamarin River this line of reef offers exceptional visibility and a wide variety of fauna, coral and shells in depths ranging from twenty feet on the Harmony Reef to 100 feet and more off the main coastal reef. Boats can be hired at either Black River village or the Le Morne Hotel.

Two local clubs, the Mauritius Underwater Group and the H.M.S. *Mauritius* Sub-Aqua Club are always available to advise those who wish to venture into deeper water; it is wise to consult their experts if, for instance, one intends diving off the east coast. A constant westward current can make things extremely difficult for those inadequately equipped.

Chapter 18

Diving off the Seychelles

A rare undersea experience awaits those who travel across the Indian Ocean to the Seychelles Islands, for until recently only a privileged few were able to undertake the journey. The islands include Aldabra, the largest coral atoll in the world, as well as the Amirante and Farquhar groups of islands. Unlike the islands in the Caribbean, these Indian Ocean outposts are only just being discovered by the diving fraternity.

'More of sea than land' the early French privateers often referred to a vast expanse of island-speckled ocean straddling one of the major sea routes to the East, just south of the Equator and just north of Madagascar.

They were right too, for while the ninety-two islands which comprise the Seychelles group are situated in an expanse of Indian Ocean as large as Spain and France together, the total land-mass of all the islands together would barely cover an area as large as Greater London.

But it was the beauty – breathtaking and stuck away in a remote corner of the world – which attracted the occasional early traveller as it lures those who can afford it today. They came then as they arrive today, even though a holiday in the Seychelles less than a decade ago was still measured in months if one was travelling by sea from Europe, for there was no other way. At the present time it takes fourteen hours by jet from Europe.

Partly because of her geographical isolation and an element of mystique usually associated with remote islands – travellers thought twice before going there – so the Seychelles Islands have survived with much of their primaeval lore intact. In an age when polution and neglect is taking a ready toll in former hideaways such as the Caribbean and Eastern Mediterranean, the Seychelles remain cloistered by distance; a legacy of yesterday for those of tomorrow and it is exactly because of this that the islands are regarded by divers as among the most fascinating to be found anywhere.

What makes it all the more interesting to underwater *cognoscenti* is

In the Seychelles Pauline
Saw of Malindi, Kenya, dives
past a group of parrot fish.

the variety. While some islands are little different from their counterparts in any other tropical waters, others, notably Aldabra, the African Banks, Amirantes, L'llot (Fregate) Island, among others, portray undersea vistas which in many instances defy description. Cosmoledo Island is world renowned for its marine turtles which, on account of fairly stringent conservation controls, remain impervious to the presence of divers. Aldabra is little short of the diving experience of a lifetime to anyone who has been there, and there have not been many.

Even in some of the more accessible regions such as Mahé – the largest island of the group, seventeen miles long and between three and five miles wide – this relatively tiny area boasts a tapestry of sandy beaches and palm-fringed coves almost unequalled in any

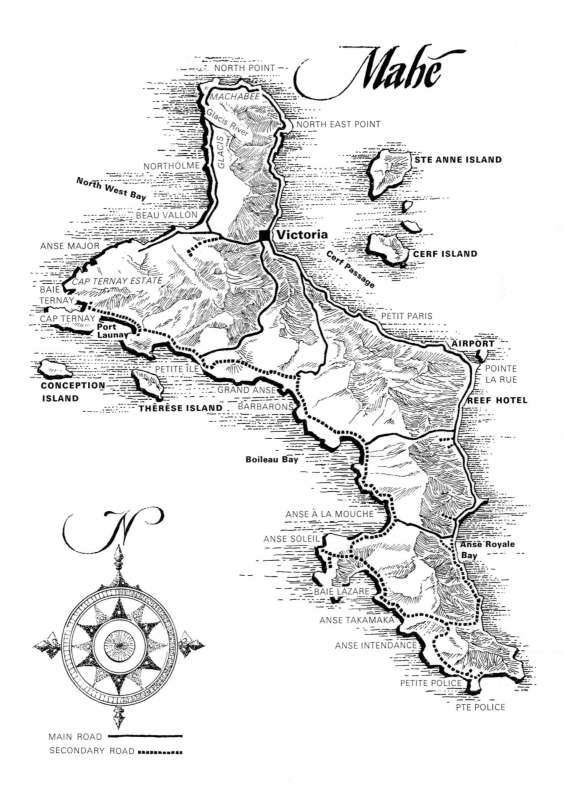

Mahé

NORTH POINT

MACHABEE

Glacis River

NORTH EAST POINT

GLACIS

NORTHOLME

STE ANNE ISLAND

North West Bay

BEAU VALLON

Victoria

CERF ISLAND

ANSE MAJOR

Cerf Passage

CAP TERNAY ESTATE

BAIE TERNAY

PETIT PARIS

CAP TERNAY

AIRPORT

Port Launay

POINTE LA RUE

PETITE ÎLE

CONCEPTION ISLAND

GRAND ANSE

REEF HOTEL

THÉRÈSE ISLAND

BARBARONS

Boileau Bay

ANSE À LA MOUCHE

ANSE SOLEIL

Anse Royale Bay

N

BAIE LAZARE

ANSE TAKAMAKA

ANSE INTENDANCE

PETITE POLICE

PTE POLICE

MAIN ROAD ━━━━━
SECONDARY ROAD ▪▪▪▪▪▪▪

ocean. The island rises a spectacular 3 000 feet out of the sea and in places granite peaks are interspersed with a variety of lush forests; contrasts are stark but inviting.

Offshore; coral, visibility, porpoises, the occasional shark (although there are not many) and every size and hue of tropical fish are there for the viewing; one has only to know where. These include anything from tiny multi-coloured wrasse all the way through the scale of blue-and-yellow striped surgeons, *Acanthrus lineatus*, bat rays *Stroasodon narinari*, mantas, black-and-white butterfly fish *Hemitaurichthys zoster*, tunny, barracuda and, if you are lucky, the magnificent ocean-blue sailfish.

Diving facilities on Mahé island have improved markedly since the airport was opened early 1972. From conditions bordering on the primitive, visitors are today able to purchase the full range of diving gear from a variety of stores in Victoria, the main town. Spares are available for some of the more popular American and European brand names.

Port Victoria, Seychelles, is where most things happen in the British Indian Ocean territories. (Peter Saw)

Scuba facilities too are on the market for hire. A number of firms have their own compressors to fill your bottles. Everything needed for a diving expedition has to be taken if the visit is extended to Praslin and other islands.

Some of the finest diving is done from boats; these too are available but their hire can be expensive if groups of divers are not prepared to club together for joint underwater sessions.

A day-long diving trip to Denis or Bird Islands, for example, where conditions rank with the best available, can cost as much as £80 for a maximum of ten persons. If food and drinks are included the price can be increased by £20. Other, shorter trips can easily be arranged for parties from about £40. Again, it is a question of similarly-interested individuals prepared to hike along and split costs. More boats are planned for the future and it is likely that costs will be lowered as competition increases. It is advised that all prospective divers get as many quotes as possible from boatmen in Victoria harbour.

While diving off the Seychelles is first rate by European standards, there are areas where the visibility ranges from fair to poor – particularly along the eastern fringe of Mahé Island. The entire thirteen-mile stretch southwards from North East Point, through Cerf Passage, past Point la Rue and South East Island and farther south to Anse Royal Bay is comprised of a shallow studded reef with coral banks and low ridges. The visibility here is not the best and in many instances the coral has been blasted by dynamite (since banned) but is slowly reaching maturity again. For conchologists, small fish study and close-up photography in this portion of exposed sea is excellent.

The main diving therefore, starts just north of South West Point, around the bulge and all the way down the west coast. Each area is dealt with in turn.

North East Point to North Point This area is rocky. Because of constant north-east swell some caution close inshore has to be observed. Visibility here averages about thirty feet. Fish are in abundance, particularly game fish in deeper waters.

Machabee (off North Point) One of the best areas for seeing a variety of fish – including game fish, 'cuda, grouper, porpoise. Visibility improves all the while from the murky areas farther to the south-east. Large coral heads off Northolme are worth investigating.

Glacis   The scenery in this area of the island is extremely pretty. Underwater visibility is good but not all that interesting, mainly because of a sandy bottom. There is not nearly as much marine life as farther north and towards the south-west.

North West Bay (Beau Vallon)   A number of boats set out from this area to dive elsewhere. Conditions are not unlike those of the Glacis River area.

Cape Ternay and Baie Ternay   With beaches ringed by palms and rocks this area is one of the most attractive sites on the island. Together with Port Launay it forms one of the marine parks in the Seychelles. Because the visibility in this area often reaches 100 feet it is also one of the best areas for diving. Fish of all sizes are found, including angel, butterfly, sergeant-major, green humped-head and parrot fish in vast numbers, interspersed with larger game fish. On the edge of the bay it is interesting to dive between large granite slabs which have lain there since primeval geological action originally formed the islands. Coral starts improving markedly compared with the north.

Conception and Thérèse Islands   An area of lovely rocks and undersea caves with conditions ideal for diving, most interesting. It is necessary to dive in inshore areas because there is no coral on the seaward side. There are numerous kinds of parrot fish. Shells abound, including helmets, tritons and conches. Thérèse Island is private property and a permit is necessary to dive here. Visibility in the vicinity of Petite Île is good.

Barbarons   Moving farther towards the east down the coast one reaches one of the best diving spots off Mahé at Barbarons. Visibility is usually at least 100 feet, and there are fish in abundance but nothing big. Again, like Conception, there are many large granite rocks covering the undersea terrain.

Anse à la Mouche and Anse Soleil   There is not too much interest generally, as it is shallow and murky, but it is worth moving a short distance to Anse Soleil where visibility is between eighty and sixty feet. This area is known for the occasional shark, particularly because it lies close to one of the shark-breeding zones in the Seychelles. Caution should be exercised although there is no record of attacks in recent years.

Baie Lazare   A shallow area with fair diving, but it is not the most exciting although good for beginners and offering a variety of dips in the protected bay. The coral here is average. No problems with sharks are recorded.

Anse Takamaka A sandy bottom prevails with visibility about eighty feet. Shells are readily seen along the sea bed. On the southern side divers will find a rocky ledge which has some interesting marine life.

Anse Intendance Although this zone also has a sandy bottom, this is an interesting underwater stretch with fair visibility. The surrounding land area is beautiful and the beach is ideal for a day's outing and picnics. There is no coral to speak about. The southern tip of the bay is clustered with numerous varieties of smaller fish.

Petite Police Considered by many as the best diving spot off Mahé, visibility is excellent – about 100 feet plus for much of the year. The water reaches a depth of sixty feet just off the southern peninsula. A fair-sized colony of large groupers live in the area, apart from most types of game fish, including barracuda, tunny and sailfish. Police Point to the east is ideal for the study of fish life, but the area east of the lighthouse is a private plantation so it is not possible to dive there unless permission is granted.

Praslin The ferry boat *Lady Esmé* leaves Port Victoria three mornings a week for the round trip, Praslin, La Digue and back to Mahé. The charge is reasonable though it takes nearly five hours to cover thirty miles to Praslin and back.

Praslin sunset. (Peter Saw)

With an hotel on the island, facilities for diving are ideal, although at the time of writing there were no facilities for recharging scuba equipment. Generally sea conditions around Praslin are better than on Mahé. Point Rouge, a sandy-coloured boulder-strewn point at the southern tip of Curieuse Island, just off Praslin, offers some of the best underwater viewing possibilities. Visibility is good and depth off the point is roughly forty feet. Divers will find schools of large bat rays, known locally as *Raie Chauve-Souris*, and named after the island's flying foxes or fruit bats. The newcomer is likely to be awed by the *Carangue Chasseur* which surrounds the diver in a whirlwind of silvery-grey soon after entering the water. Angel fish, buttercup-yellow fish with their brilliant-blue lips, butterfly fish and a variety of others are found here as are the green hump-head wrasse, which are wary of visitors.

Sharks are also active in this area but they rarely trouble visitors. They are interesting if only for purposes of observation as clusters are found in the deep gulleys from time to time. Most swim in lazy circles and then disappear into the distant murk. A number are of the family *Carcharhinidae*, more commonly known as 'greys'. Few measure more than seven feet.

Pauline Saw, who often dives in the Seychelles, recounted one experience off Praslin: 'We moved the boat along and re-entered the water. Here we found the sharks! Below me, at the edge of the visibility, a long sinuous shape moved above the sand and quickly disappeared down the coral gutter. Swimming on, a number of these ominous shapes became a weaving group, seemingly performing a ritualistic dance. Peter, my husband, dived towards them, while I tried to keep right out of the way. Peter counted eight and then lost count as they swam lazy circles appearing and disappearing into the blue.

'They were in no way aggressive; as Peter dived towards them with his camera. they slid away. The water clarity was not too good as the plankton fogged the water. However, again and again Peter dived, more or less playing "tag" in his endeavour to record the event on film.'

There is an interesting spot in the channel between Côte d'Or and St. Pierre, but here plankton can make things uncomfortable for some and photography impossible. One of the attractions are the large Manta rays which swoop past on occasion, and you may also be able to spot a school of *Diable de Mer*, *Mobula diabolus*. Along the side of the channel are low, damaged banks of coral with an occasional large coral head sticking through. There is evidence that the crown-of-thorns starfish, *Acanthaster planci*, is active in this area.

Rock cod appear to be scarce in these parts, although octopi can be spotted by those interested enough to search. They are difficult to recognize unless you know what to look for as conditions for camouflage are ideal.

La Digue Island Lying to the east of Praslin and easily reached from Victoria, there is good diving on the northern side with visibility up to 100 feet.

The St. Anne Marine National Park Excellent diving can be enjoyed in this area to the north-east of Port Victoria with a visibility of about sixty feet, with numerous coral banks. Because this is a sanctuary no marine life including shells may be touched. Cerf Island is good for photography with excellent coral. Again nothing may be touched or removed, and this warning should be heeded as conservation controls are stringent.

Denis and Bird Islands These two spots on the map about sixty miles from Port Victoria are the only coral islands in the Mahé area, and diving in their vicinity is outstanding. They are world-renowned for their bird life – Bird Island is the nesting site of the sooty tern. Underwater visibility off both islands is in excess of 100 feet. Although they are reached by fast boat from Mahé, there is an airstrip on Bird Island. Accommodation is available in a number of fishing lodges which provide 'self-help' facilities.

Amirantes Lying to the south-west of the Seychelles proper, this unspoilt, virtually untouched cluster of tropical islands, like Aldabra, is one of the best and most interesting diving spots anywhere. Outstanding for undersea beauty is St. Joseph's Island, which is the largest and lies in the middle of the group. Poivre is privately owned. Underwater visibility in the group ranges between 200 and 300 feet. There are few facilities on the Amirantes; no hotels, no airstrip, no petrol and few shops in the accepted sense of the word. Sea life in the vicinity is fantastic with game fish, morays, shells and turtles. There are also sharks, though with fine visibility they can be spotted from a distance, so offer no problem. A schooner with passenger accommodation onboard leaves Port Victoria at regular intervals for the Amirantes; the round trip usually takes a week and charges are reasonable.

Index

Index